AI TODAY: OI TOMORROW

THE DAWN OF ORGANOID INTELLIGENCE: OPPORTUNITIES, ETHICS, AND THE FUTURE OF HUMANITY

EDMUND J.B. WHITE

ISBN: 978-1-950043-58-3

Publishing services provided by Archangel Ink

Illustrations
The author conveys appreciation for the permission to use the images featured in this book. Every effort has been made to communicate with all copyright owners, and any discrepancies brought to the author's notice will be rectified in forthcoming editions.

Endless Constellation background provided by SVGBackgrounds.com

Book Cover
The author would like to thank Mike Trent for the imagery used for the front cover.

Learn more about Edmund (Ed) J.B. White on LinkedIn or visit ai2oi.com to stay updated on his latest insights, projects, and discussions about the future of technology and humanity.

Edmund J.B. White | LinkedIn

DISCLAIMER

The information presented in this book is for educational and informational purposes only and is current as of the publication date. While every effort has been made to ensure the accuracy and completeness of the content, the author and publisher make no representations or warranties of any kind, express or implied, regarding the accuracy, reliability, or completeness of the information provided.

The content is not intended to substitute for professional advice, whether medical, legal, financial, or otherwise. Readers should consult appropriate professionals for advice on specific situations. The author and publisher disclaim any liability to the fullest extent permitted by law arising directly or indirectly from the use of this book, including but not limited to any errors or omissions contained herein.

Any views or opinions expressed in this book are solely those of the author and do not represent those of any organizations or institutions with which the author may be affiliated. References to products, services, or organizations are for informational purposes and do not constitute an endorsement

The author acknowledges that the term "organoid intelligence" (OI) is subject to ongoing debate within the scientific community. Some experts have expressed concerns that this term may be misleading or non-scientific, and do not endorse its use. The author's decision to use "organoid intelligence" in this book reflects the currently accessible perspective on the potential of this emerging field and is intended to explore its opportunities, ethical considerations, and future implications. However, the author remains open to changing this perspective as new research emerges and the scientific community continues to discuss and refine the terminology surrounding this field. Readers should be aware that the term "organoid intelligence" is not universally accepted and are encouraged to consult a variety of expert sources to form a well-rounded understanding of the debate surrounding this terminology. The author's views on this matter may evolve over time, and he welcomes readers to join him in this ongoing intellectual journey.

The author and publisher shall not be liable for any damages or losses arising from the use of information contained in this book. Readers should not attempt to replicate any experiments or procedures described herein without proper training, equipment, and precautions. All trademarks, service marks, product names, and logos appearing in this book are the property of their respective owners.

For my mother, Noreen White, whose love knew no bounds.

CONTENTS

PART IV: Playing God or Serving Humanity? The Ethics of OI

PART V: Charting the Course for Responsible OI

INTRODUCTION

THE DAWN OF OI: HOW BRAIN ORGANOIDS ARE REDEFINING WHAT IT MEANS TO THINK

Picture this: a layer of living brain cells, no bigger than a fingernail, wired up inside a lab and given control of the classic video game Pong.[1] At first, the neurons flail helplessly as the digital ball whizzes by. But with a little feedback and a lot of practice, something extraordinary happens—the cells start to learn. The Pong paddles slide into place with greater precision. The neural network adapts and improves. Soon, this microscopic sheet of brain tissue is deftly batting the ball back and forth, mastering a crude version of one of the earliest computer games.

If you think this sounds like science fiction, you're in good company. But a surreal breakthrough actually happened in 2022. A team at Cortical Labs in Melbourne, Australia, grew networks of human neurons in a dish and showed they could learn to play Pong in real time. Wired up to electrodes and provided feedback on their performance, these living neural networks adapted their firing patterns to gradually improve at the game. It was a pivotal "world's first" demonstration that flipped the script on what lab-grown neurons can achieve. This groundbreaking research has given rise to a new field called "organoid intelligence" (OI), which explores the computational power of lab-grown brain tissue.[2] While artificial intelligence (AI) aims to reverse engineer human cognition in silicon, OI offers the tantalizing prospect of replicating the brain's raw processing power more directly in living neurons.

For simplicity's sake, in this book we'll be using the term "brain organoids" to refer to several different types, including unguided brain organoids that develop spontaneously from stem cells, regionalized brain organoids guided to mature into specific brain regions, assembloids that combine different organoid types, and grafted organoids transplanted into animal models.

Throughout this book, we'll explore the rapid advancements in brain organoid research and the potential applications of OI, from ultra-efficient biocomputers to powerful tools for simulating neurological diseases. We'll also grapple with the profound philosophical questions this technology raises about the nature of intelligence and what it means to be human.

While we'll be focusing on the key concepts and applications of brain organoids, it's important to recognize the precise terminology used in the field. If you're interested in diving deeper into the technical classifications, I highly recommend checking out the scientific paper "A nomenclature consensus for nervous system organoids and assembloids" published in the journal *Nature* in 2022.[3]

As we embark on this journey, it's crucial to keep in mind that even the most advanced brain organoids today are still simplified models compared to a real human brain. However, they are already proving to be invaluable in the quest to unravel the mysteries of the brain and unlock the potential of OI.

Now consider a more concerning scenario: a totalitarian state furtively collecting tissue samples from its citizens to develop brain organoids in pursuit of a new form of biological computing. Imagine thousands of brain organoids pulsing with electrical activity and wired into a classified supercomputer for the purpose of surveillance, cyberattacks, or advanced weapons research. In the wrong hands, brain organoid technology could be a powerful tool for oppression and conflict. An artificial superintelligence, built from an interconnected web of human brain organoids, unleashed to dominate the digital realm and impose the will of a tyrant. It may sound like dystopian science fiction, but it could be a plausible threat in a future where organoid research advances unchecked.

In this book, we'll take you on a journey into the fast-developing world of brain organoids and "organoid intelligence." We'll trace the field's rapid progression, from discovering how to grow brain organoids, to teaching them to perform simple tasks similar to computers. While still early, these advances hint at the mind-bending future possibilities of biological computing and brain-machine interfaces. We'll explore the potential implications of this groundbreaking technology and what it could mean for the future of computing and our understanding of the human brain.

You'll learn how scientists are culturing ever more sophisticated miniature brain models, and you'll meet the ambitious researchers, bioengineers, and visionaries working to make OI a reality. Along the way, we'll grapple with the profound philosophical questions this technology raises: Can we really capture the dynamics of human cognition in a dish? What are the ethics of creating "living" computers? And if we can alter the physical basis of thought, how might that reframe what it means to be intelligent—or even human?

Whether you're excited, intrigued, or unsettled by the notion of brain organoids that can think and learn, this book will deepen your understanding of the very nature of intelligence and open your eyes to how the strange new science of OI—today's successor to AI—may soon reshape it. Strap in. The era of organoid intelligence is here, and you're about to see how it could change everything.

CHAPTER OVERVIEW

This book endeavors to provide you with an engaging exploration of OI, divided into five parts. Part I kicks off with an introduction, laying the foundation of OI and its incredible potential. In Part II, we start journeying through the revolutionary applications of OI across diverse fields, from medicine to space exploration, and discover how it could transform our daily lives. Part III takes a cautionary turn, exposing the potential risks and challenges that come with this groundbreaking technology. Part IV confronts the ethical dilemmas head-on, asking whether OI is a leap forward for humanity or a step too far. Finally, Part V brings it all together, proposing a framework for responsible OI development and regulation, ensuring that this powerful tool benefits society while minimizing risks.

Part I: Unveiling the Mysteries of Organoid Intelligence (OI)

Chapter 1: Unveiling Brain Organoid Basics

This chapter introduces brain organoids, miniature structures grown from stem cells that mimic aspects of the human brain. It details the meticulous process of creating these organoids and explores their capabilities and limitations as simplified brain models.

Chapter 2: Realizing OI Potential

This chapter delves into the challenges of developing functional OI, which requires mastering organoid engineering to create larger, longer-lasting structures that closely mimic the human brain. Key focus areas include building high-tech interfaces for precise neural recording and stimulation, providing organoids with sensory inputs and motor outputs to facilitate learning, and establishing high-speed data networks and adaptive algorithms to seamlessly integrate organoid and AI cognition.

Chapter 3: Neural Nurseries

This chapter explores the potential of brain organoids to help our understanding of cognition and AI. By studying these organoids, scientists aim to bridge the gap between biological and artificial intelligence, unlocking insights into learning, memory, and adaptation. Key challenges include enabling organoids to learn from minimal data the way humans do and creating multisensory environments for more holistic learning. This chapter also addresses the technical difficulties and starts exploring some of the ethical hurdles in developing intelligent organoids.

Chapter 4: Synapse and Silicon: Decoding Neural Networks

This chapter explores the exciting potential of combining advanced computational techniques like artificial intelligence (AI) with brain organoids to gain deeper insights into neural networks and brain function. It describes how AI algorithms can interpret the complex biological data from brain organoids, accelerating the pace of discovery. The chapter highlights technologies like "brain-organoids-on-a-chip," closed-loop discovery algorithms, and the evolution of the "DishBrain" system into the more advanced CL-1 device, which enables a higher level of control and analysis of brain organoids. However, it also emphasizes the continued importance of human insight to guide the AI and the need to avoid hype and have realistic expectations about current capabilities. It touches on the future potential of quantum computing, with its ability to leverage principles like superposition and entanglement, to further enhance our understanding of brain organoids and neural networks.

Part II: OI – Revolutionizing Life as We Know It

Chapter 5: Insights from Lab-Grown Minds

This chapter introduces the impact that brain organoids have on the understanding of brain development and disorders. While not full replicas, these brain organoids recreate key aspects of brain biology, offering valuable insights into neurobiology that bridge lab research and real-world biology. Scientists use brain organoids to study the brain's intricate folds and layers, essential for cognition, and to explore neurological disorders by modeling disease progression. These breakthroughs promise to revolutionize medicine, from drug discovery to personalized treatments, though ethical questions remain.

Chapter 6: Medical Revolution: Dish-Derived Healing

This chapter delves further into how brain organoids could transform medicine by offering more accurate models for drug discovery and disease research. Traditional animal models often fall short in predicting human drug responses, leading to costly failures. In contrast, brain organoids, grown from human stem cells, replicate key structures of the human brain, enabling more efficient drug screening and reducing development time and costs.

Brain organoids also pave the way for precision medicine, allowing patient-specific models to predict individual treatment responses. They provide new insights into disorders like Alzheimer's and autism, helping identify therapeutic targets. While challenges remain in scaling and refining the technology, brain organoids hold great potential to transform medicine and tackle complex diseases.

Chapter 7: Neurological Clarity: Role of Lab Neurons

This chapter does a deep dive into how brain organoids are transforming our understanding of neurodegenerative and psychiatric disorders to uncover the causes of conditions like Alzheimer's, schizophrenia, autism, depression, and anxiety. In Alzheimer's research, patient-derived organoids replicate disease features such as amyloid plaques, offering insights and testing grounds for treatments. For schizophrenia, organoids reveal neural misfiring and gene-expression abnormalities, guiding targeted therapies. Autism research benefits from these models, exposing developmental imbalances and paving the way for personalized interventions. Though depression and anxiety research is early, organoids offer new avenues to study neurotransmitter imbalances. The chapter also explores the gut-brain connection and the future potential of brain organoids in restorative medicine to repair neural circuits in neurodegenerative diseases.

Chapter 8: Memory in Brain Organoids: Fact or Fiction?

This chapter explores the potential for brain organoids to develop memory, raising questions about their ability to imagine, learn, and form a sense of self. Memory is framed as crucial to intelligence, learning, and identity. Ongoing research could enable organoids to manage complex memory tasks, with applications in learning and problem-solving. The chapter also questions whether memory alone could lead to consciousness, highlighting the challenge of replicating human-like memory without real-world experiences.

Chapter 9: Brain Organoids: Unveiling Sleep and Dreams

This chapter explores the possibility that brain organoids might experience sleep and dream-like states as they grow more complex. It highlights sleep's essential roles in memory consolidation, metabolism, and synaptic remodeling, and suggests that the neural activity seen in organoids could mimic basic sleep patterns. Organoids may help reveal why sleep is crucial, even in lab-grown brains, especially if enhanced with artificial bodies and environments to establish realistic sleep–wake cycles. The chapter also touches on the effects of psychedelics on organoids, suggesting they possess the circuitry for complex brain functions, offering insights into consciousness.

Chapter 10: Brain Organoids in Space: A New Frontier

This chapter explores how brain organoids are being used to study the effects of space travel on the human brain. These models offer insights into how microgravity and cosmic radiation impact brain development, with projects like the BOARDS (Brain Organoid Advanced Research Developed in Space) mission by UC San Diego and Space Tango sending organoids into orbit to study brain function in weightlessness.

It highlights the challenges of keeping organoids viable in space and explores biohybrid interfaces for spacecraft systems. Ethical concerns about organoid use in space are raised, stressing the need for fairness and humanity.

Part III: The Potential Perils of OI

Chapter 11: Lab Life: The Consciousness Debate

This chapter explores the profound questions surrounding the potential for consciousness in lab-grown brain organoids. Consideration is given to diverse theories on the origins of consciousness and the implications if brain organoids were to develop awareness, including impacts on research practices and moral obligations. The chapter also examines the cutting-edge technique of xenotransplantation—introducing human brain organoids into animal brains—and its promise and perils.

Chapter 12: OI Biosecurity Concerns

This chapter addresses biosecurity concerns surrounding brain organoids, calling for proactive measures to prevent misuse. It underscores the need for strong safeguards, oversight, and ethical guidelines to ensure the responsible development of OI technology. By examining potential misuse scenarios, the chapter advocates for a balanced approach that supports scientific progress while prioritizing ethical OI development. It emphasizes the importance of ongoing dialogue, collaboration, and public engagement to responsibly navigate the ethical challenges of this emerging field.

Chapter 13: OI: A Sustainable Future

This chapter explores how brain organoids could inspire the development of energy-efficient, sustainable computing systems. It highlights the extraordinary computational power and efficiency of the human brain, which far surpasses that of today's supercomputers. Through a discussion of recent experiments involving organoid neural networks and the integration of brain organoids with AI to create "Brainoware" biocomputers, the chapter examines how these systems could address the high energy consumption of current AI by replicating the brain's capabilities. Despite challenges, it emphasizes the potential for brain-inspired computing to reduce the environmental impact of data processing and AI, paving the way for a more sustainable technological future.

Part IV: Playing God or Serving Humanity? The Ethics of OI

Chapter 14: The Ethics of OI: Balancing Promise and Peril

This chapter delves into the ethical complexities surrounding the development of OI and its potential to enhance human cognition. It addresses the challenges of engineering morality and empathy into OI systems, as well as the risks of runaway dynamics, where self-improving OI could quickly exceed human control and comprehension. Emphasizing a gradual approach to OI development, the chapter calls for careful oversight and collaboration among scientists, ethicists, and the public. It also explores the concept of the singularity—a hypothetical point where AI vastly surpasses human intelligence—and its profound implications for consciousness and the future of humanity.

Chapter 15: Brain Organoids and the Soul

This chapter examines how brain organoid research challenges traditional religious views on the soul and consciousness. Through dialogues with spiritual leaders, it reveals the tensions between materialist and spiritual interpretations of the mind, as these brain organoids could in the future blur the lines between life and sentience. The chapter demonstrates how religious perspectives significantly influence the trajectory of brain organoid science, from funding decisions to regulatory frameworks. It advocates for approaching this research with wisdom, humility, and reverence, positioning faith communities as essential partners in guiding its ethical development.

Chapter 16: OI and Media Influence

This chapter examines the significant role media narratives play in shaping public perceptions, expectations, and the direction of OI research. From sensationalized journalism that exaggerates potential applications to dystopian films and thought-provoking science fiction, media portrayals of OI strongly influence research priorities, funding, regulations, and public opinion. However, these depictions often lean towards hype and overlook the complex scientific and ethical issues at hand. The chapter advocates for responsible, balanced storytelling from journalists, creators, and researchers to counter sensationalism.

Chapter 17: Ancient Wisdom – Modern Minds

This chapter explores how ancient wisdom traditions from Indigenous cultures worldwide can deepen and enrich our understanding of OI. Philosophies such as Aboriginal Dreamtime and Inuit

Qaujimajatuqangit emphasize knowledge as interconnected, contextual, and rooted in human values. Applying these perspectives to OI encourages a more responsible, long-term approach, mindful of the consequences for future generations, as reflected in the Haudenosaunee Seven Generations Principle.

Chapter 18: OI in Art and Imagination

This chapter examines the fusion of art and science in brain organoid research, highlighting how artists are engaging with the profound implications of this technology. From musicians composing cellular symphonies to painters exploring the mysteries of consciousness, creative minds are pushing the boundaries of expression by drawing on advancements in OI. These artistic explorations serve as compelling metaphors, igniting discussions about the nature of intelligence, the ethics of engineering consciousness, and the essence of being human as the line between biology and technology blurs.

Part V: Charting the Course for Responsible OI

Chapter 19: Prudent Regulations in the OI Era

This chapter emphasizes the need for proactive, adaptive, and ethically grounded regulations to guide the rapid advancements in OI technology. Drawing lessons from AI governance challenges, the OI community must establish clear guidelines and oversight mechanisms to protect against misuse, ensure equitable access, and navigate complex ethical questions. Key focus areas include non-maleficence protections, the balancing of intellectual property rights with open innovation, and implementation of security measures that mitigate risks without stifling progress. Engaging diverse stakeholders in ongoing dialogue is crucial for steering OI development towards a responsible, inclusive, and beneficial future.

PART I

UNVEILING THE MYSTERIES OF ORGANOID INTELLIGENCE (OI)

CHAPTER 1
UNVEILING BRAIN ORGANOID BASICS

In labs around the world, tiny clumps of tissue are growing in specialized vessels. Those strange structures are brain organoids, which are miniature versions of human organs grown by scientists in the lab.

Organoids are created using stem cells, which are special cells that can develop into any type of cell in the body.[4] Researchers carefully guide these stem cells to grow and organize themselves into tiny, simplified versions of organs like the brain, heart, or liver.

These "organs-in-a-dish" are usually only a few millimeters in size, but they mimic the basic structure and function of their full-sized counterparts. While they aren't exact replicas, organoids provide a unique window into how our organs develop and function on a microscopic level.

These quirky blobs floating in culture may not seem very impressive or "brainy" upon first glance. But looks can be deceiving. Just as Einstein's eccentric appearance hid his brilliance, the unassuming organoid hides its power within. For under the microscope, these tiny tissue creations come to life: neural networks firing, intestines secreting, lungs breathing. Tiny organs, but with plenty of swagger.

What could possess someone to want to grow a brain, intestine, or liver in a laboratory? Is this some mad science experiment or a genuine health breakthrough in the making? Read on and let your inner skeptic be amazed as we explore how bioengineers are unlocking the secrets of nature and human development, using organoids as their key: crafting living models where neuroscientists can peer into the enigmatic events sculpting the fetal brain, oncologists can witness tumors developing in a dish, and pharma researchers can test drugs on human tissue without leaving a mark.

But appreciating the significance of organoids requires first understanding the method behind the madness. How do scientists make them in the first place? What exactly are they mimicking, and what are their limits? Grab your lab coat and let's dive deeper into the peculiar process of creating an organoid.

The Meticulous Science of Manufacturing Miniature Organs

Growing a miniature-sized version of the brain, liver, or intestines in a specialized vessel is no cakewalk. It entails guiding stem cells through a series of steps aimed at getting them to organize into a 3D structure approximating the simple elegance of real organs.[5] This process builds upon our growing understanding of the factors influencing stem cell fate and behavior. It's like choreographing a dance among stem cells. The step-by-step protocol to transform stem cells into living, breathing organoids involves some serious science with a dash of artful improvisation. Let's break it down:

Step 1: Procure the Cellular Raw Materials

The first step is getting your hands on stem cells, biological building blocks that can multiply and potentially turn into any cell type. Imagine stem cells as brand-new LEGO sets waiting to be assembled into either a Star Wars ship or Hogwarts Castle, depending on the instructions provided.

Scientists obtain these shape-shifting cells either from very early embryos or by genetically reprogramming mature cells back into stem cells. Think about how a caterpillar's destiny is transformed when it undergoes metamorphosis into a butterfly capable of flight. Similarly, by adding genes like Oct4 and Sox2, even a fully formed skin cell can have its identity stripped away and reverted into an unspecialized stem cell primed for fresh instruction.

With stem cell populations prepped, scientists next begin genetically nudging them towards an organ lineage.

Step 2: Steer Stem Cells into Organ Cell Types

Scientists don't merely hope that stem cells spontaneously drift towards becoming intestinal or kidney cells. They deliberately steer stem cells into becoming specific organs.

They accomplish this by concocting cocktails of growth factors—proteins that influence development—and adding them to the cellular mixture at just the right times. These are the musical cues that direct the stem cells' dance moves.[6] For instance, BMP and FGF signaling proteins induce the first waltz steps of stem cells differentiating into intestinal progenitors.

Timing is everything. Flick that neural growth switch on too soon or too late and you'll birth some freakish mutant organ. Luckily, scientists cleverly build on decades of developmental biology knowledge to help avoid this.

With cells successfully gliding towards becoming a particular organ, the next step is initiating self-assembly.

Step 3: Coax Cells into Clumping into Organoid Blobs

Left alone, stem cells tend to just spread out lackadaisically across the petri dish. Stem cells won't develop into organs on their own; they need to clump together—or *condense*—into three-dimensional arrangements to form organoids.[7] Inspired by embryology, scientists have devised some crafty tricks to get cells to aggregate.

One way is to place cells into tiny droplets where they are forced to snuggle up. Another is to culture cells on spinning platforms that prevent surface-sticking and encourage condensation into spheroid blobs. Think about how water droplets fuse when rolled together; cells do much the same, merging into larger tissue aggregates destined to become organoids.

Step 4: Let Organoids Mature in Bioreactors

Now the cellular blobs require weeks to months of nurturing to develop into proper organoids—like caring for a chick before it fully hatches. This critical maturation phase takes place in spinning bioreactors, where organoids are perpetually suspended while slowly twirling to enhance nutrient absorption and waste removal.[8]

Other bioreactors called microfluidics chips allow ultra-precise control of the cellular environments using networks of fluid channels and chambers.[9] Imagine organoids comfortably nested into biosynthetic trees receiving water from an intricate irrigation system.

Over time, organoids morph from crude balls of cells into defined mini-organs exhibiting some of the grace and functionality of the real thing.

Step 5: Keep Organoids Alive and Thriving

Sadly, stem cells can't just be left alone in a warm incubator to grow indefinitely. There are challenges to keeping organoids viable long-term.

One specific challenge is that, as the organoids grow, their need for oxygen and nutrients becomes greater than the amount that diffuses in. It's like forcing someone to live in a house without doors or windows: pretty soon they suffer if there's no air getting inside.

To prevent the untimely demise of their precious organoid creations, scientists are trying all sorts of tricks: installing networks of blood vessel–like tubes to ferry oxygen, optimization of media changes, and integrating microfluidic life-support systems. The ultimate goal is culturing happy, healthy organoids for months and, hopefully, for years.

And voila. With plenty of patience and skill, crude aggregates transform into living, breathing organs-in-miniature. But how close are they to the real thing? Let's examine more closely what organoids can and can't do.

What Exactly Can Brain Organoids Mimic? Peering Into the Powers and Limits of Miniature Organs

Never judge a book by its cover. Likewise, don't judge a brain organoid by its small size or blobby appearance, for within even a pinhead-sized organ lies astonishing cellular complexity. Scientists are continuously uncovering new details about what these models can achieve.

Structure: The Inherent Beauty of Miniature Organs

Let's start with what brain organoids can recreate relatively well: the innate three-dimensional architecture of brains. Unlike boring flat cells growing on plastic, brain organoids blossom into spheres flecked with intricate cellular landscapes that reflect real tissue forms.

The intrinsic ability of stem cells to self-organize into lifelike tissue structures is perhaps the most profound aspect captured by organoids.

Composition: Concocting Simplified Cellular Cocktails

Peer closer and you'll see that most brain organoids display a diversity of cells like those found in the organ they are imitating, just fewer in number and type. For example, brain organoids contain neural stars like neurons, supportive astrocytes, and insulating oligodendrocytes, key ingredients of the central nervous system.[10] While far short of the staggering multitudes of cells in real organs, brain organoids brew together enough cell varieties to enable scientists to study interactions between them and gain key insights.

But hold up, let's give some love to the glial cells! These unsung heroes, like astrocytes and oligodendrocytes, aren't just backup singers. They're the new rock stars of the brain, taking center stage in keeping our noggins healthy and humming. Glial cells are social butterflies, helping neurons chat and make memories. And when things go wrong, they're the first responders, working to patch things up.

In brain organoids, glial cells and neurons mix and mingle, giving scientists a front-row seat to the neural cocktail party. By watching these cellular BFFs interact, researchers can unlock secrets about how the brain works and why it sometimes goes on the fritz.

Genetics: Manifesting Patient Health in a Dish

Organoids can additionally reflect a patient's unique genetic makeup since they are brewed from a person's cells. For instance, researchers can take a skin sample from an individual with cystic fibrosis (CF), reprogram those cells into stem cells, and then generate lung organoids exhibiting that CF patient's particular disease mutations.[11]

This means organoids can manifest a wide spectrum of genetics—from the healthy to diseased—to facilitate the study of health and illness tailored to diverse people.

Functions: Rudimentary Mimicry of Organ Physiology

It's here that brain organoids display their greatest limitations. While they can recreate some simple functions of real organs, full physiological repertoire is out of reach.

Brain organoids exhibit only basic neural impulses, far short of conscious cognition. Stomach organoids secrete digestive enzymes but couldn't process your most recent meal. Genital organoids form sperm and eggs but cannot engineer a baby.

The takeaway is that organoids achieve scattered pockets of functionality that provide glimpses into normal organ processes, though nowhere near total performance. They are at best semi-competent substitutes for the real things.

Maturity: The Perpetual Infancy of Lab-Grown Organs

A primary constraint is the failure of brain organoids to reach adult maturation like real organs. Limitations in sustaining their growth artificially means they typically plateau at an immature fetal-like state.

While they progress through early aspects of development—acquiring structure, cell diversity, functionality—brain organoids hit growth restrictions before full maturity. It's like they suffer from chronic pediatric arrest. This stagnation at sometimes only an embryonic or fetal stage limits insights into late development and full-fledged disease phenotypes. It is a primary challenge curtailing their utility.

The bottom line is that brain organoids show enormous promise to recreate hallmark features of our organs in bite-sized form—enough biology to give scientists a remarkable window into aspects of human development, health, and disease that were previously out of view. Are they ready to replace animal research or become transplantable organs? Certainly not yet, but the possibilities are abundant if their remaining limitations can be overcome through continued innovation.

Concluding Thoughts

Our journey through the microscopic landscape of brain organoids has unveiled a universe within a specialized vessel.

Just like a magician revealing the secrets behind an astounding trick, we've demystified the meticulous process of creating these miniature marvels.

From the initial procurement of stem cells to the intricate dance of cells multiplying and taking shape to form brain organoids, each step in this journey is a testament to the blend of art and science that defines modern bioengineering. Brain organoids, in their tiny yet complex existence, provide a mirror to our own biological intricacies and a canvas for future medical breakthroughs.

However, as we've seen, the story of brain organoids isn't just about replicating human organs-in-miniature. These tiny structures hold the potential to revolutionize our understanding of the brain and pave the way for groundbreaking advancements in OI. In the next chapter, we will embark on an epic quest to realize the full potential of OI, a scientific Goliath that challenges the very limits of our knowledge and capabilities.

CHAPTER 2

REALIZING OI POTENTIAL

Imagine standing before a scientific Goliath, a challenge so immense that it seems insurmountable. This is the realm of developing functional OI, a goal that towers over researchers like a colossal enigma. It's a quest that demands not just knowledge and skill but a daring leap into the unknown.

In this epic battle of intellect, scientists are like David, armed not with a slingshot and stones but with the tools of stem cell biology, tissue engineering, bioelectronics, and machine learning. Each discipline offers a piece of the puzzle, a stone in the metaphorical slingshot aimed at unraveling the complexity of organoid intelligence.

The goal is as audacious as it is groundbreaking: to create biocomputers by integrating human brain cells cultured into 3D brain organoids with AI systems. This isn't just about enhancing computing power; it's about imbuing machines with a mimicry of human cognition, bridging the gap between biological and artificial intelligence.

In the last decade, researchers have made remarkable strides. They've created brain organoids that exhibit sparks of life—spontaneous neural activity and oscillations reminiscent of brainwaves. Nevertheless, it remains a Herculean task to seamlessly integrate these living, pulsing cellular structures with AI and robotics to forge intelligent biocomputers.

Turning organoid technology into more than just models of the brain requires an unprecedented collaboration across various scientific frontiers. It's a journey filled with potential breakthroughs and inevitable setbacks, each step taking us closer to what once was the realm of science fiction.

Mastering Organoid Engineering

Think of organoid engineering as a high-stakes cooking show where the main ingredient is human brain cells. The goal? To whip up a batch of organized 3D cultures called organoids. It's a bit like trying to make a gourmet meal, but instead of a recipe, you have a complex puzzle where each piece is a neuron.

Our current brain organoids are impressive, boasting up to hundreds of thousands of cells and emulating brain regions like the cortex or hippocampus. But let's face it—compared to the human brain with its 86 billion neurons, our organoids are more like a brainy appetizer than a full-course meal.

One of the big challenges is keeping the organoid's core alive. It's a bit like trying to ensure the middle of a giant cake is as perfectly baked as the edges. Without proper nutrition and oxygen, the organoid's center could starve. We need to develop the equivalent of a fine-dining experience for every cell in the organoid.

Creating brain organoids that accurately mimic the human brain isn't just about throwing cells together and hoping for the best. It's about crafting a precise neural network, with the right mix of cell types and structures. We're trying to replicate a process that took millions of years of evolution, and nature is a tough act to follow.

Longevity and Consistency: The Quest for the Everlasting Brain Organoid

Most brain organoids today can remain viable and functionally relevant for several months to over a year, although they do face challenges that can lead to a decline in function over time. What we need is the brain organoid equivalent of a fine wine, maturing beautifully with age. This means figuring out how to develop blood vessels within the organoids and maybe even adding a pinch of anti-aging magic.

And let's not forget about consistency. Currently, producing brain organoids is like baking a cake where every attempt turns out a bit different. What we need is a standardized, automated process.

Mastering the art of organoid engineering will be our gateway to unlocking the secrets of biological computing. It's about scaling up complexity, fine-tuning the organization, ensuring longevity, and producing these mini marvels with the precision of a master chef. Only by getting closer to the dizzying intricacies of the human brain can we begin to contemplate the reality of cognition through biological computing.

Building Brain Organoid Interfaces

Consider trying to eavesdrop on a conversation in a crowded room with thousands of people. That's a bit like what scientists are attempting to do with brain organoids. To tap into their potential for computation, we need high-tech interfaces capable of both feeding sensory information into these tiny blobs of tissue and deciphering the intricate patterns of electrical signaling between neurons.

Recording the activity of up to hundreds of thousands of cells across multiple organoids requires sensor arrays so dense, they make current technologies look like child's play. It's like trying to listen to every instrument in an orchestra with a single, old-fashioned ear trumpet. The challenge calls for innovation in flexible electronics, nanoscale biosensors, and advanced microscopy—tools sophisticated enough to peer into the inner workings of organoid neural circuits over extended periods without being obtrusive.

Currently, stimulating specific neuron groups within organoids is still in the early stages. Techniques like optogenetics and ultrasound have been akin to turning on a flashlight in a dark room—they illuminate a bit but don't offer the precision we need.[12] [13] What's required are new methods that can precisely target specific cell types across large organoids.

Brain organoids aren't just about electrical signals; they're also about the ebb and flow of neurotransmitters, hormones, and other signaling chemicals. It's like trying to understand the mood of a crowd by monitoring not just their words but their body language and tone. Implanting organoids with high-resolution biosensor arrays could allow us to modulate their responses through biochemical cues.

In developing these sensing, recording, and stimulation technologies, the trick is to avoid the equivalent of stepping on the organoids' toes. We need novel biomaterials and innovative neural lace devices that integrate seamlessly into the tissue, providing a bridge between the living matter and external electronics. The challenge is to find the right balance between extracting useful signals and preserving the integrity of these sensitive biological systems. It's a dance of precision and delicacy, where every step counts.

Brain Organoids: The Not-So-Spontaneous Adolescents

Our current organoids can be compared to teenagers showing off their first spontaneous sparks of activity. It's fascinating, but to develop advanced cognitive abilities, they need a bit more guidance. Organoids, unlike intact brains, don't get to experience the world. They don't have the sensory inputs and stimuli that come from interacting with their environment. It's like trying to learn a language without ever hearing it spoken.

Currently, organoids don't have behavioral outputs or goals, making their spontaneous activity a bit aimless. By interfacing them with synthetic bodies, like cyborg insects or robotic limbs, we give them motor feedback and learning objectives. Think of giving a brain a car to drive. Suddenly, there's a purpose to all that neural activity.

Neuromodulation: The Secret Sauce for Learning

Introducing neuromodulator neurotransmitters like dopamine, acetylcholine, and serotonin could be the secret sauce that facilitates learning mechanisms in organoids.[14] Combined with sensory-motor training, such neuromodulation might enhance goal-oriented cognition, memory formation, and reinforcement learning.

In essence, to realize advanced intentional intelligence in organoids, we will need to create developmental environments that replicate natural neural maturation processes. With the right environment and care, we can help these organoids blossom into something extraordinary. By treating organoids more like children to be nurtured rather than just specimens to be observed, we unlock their full potential. It's really a journey that's as much about scientific innovation as it is about understanding and replicating the complex processes of life itself.

High-Speed Data Exchange: The Neural Information Superhighway

The first hurdle is building a neural data exchange network that operates at breakneck speed. This network must handle colossal streams of sensory and brain activity data with the agility of a tightrope walker—all with minimal latency. It's the equivalent of setting up a high-speed internet connection, but instead of streaming movies, we're streaming the inner workings of a brain.

Communicating with the organoid tissue itself requires bioelectronic systems with an extremely high channel count that can operate in real time, likely wirelessly, to avoid the organoid equivalent of being tied down by phone cords.

The quest for unified cognition between biological and artificial processors calls for a new breed of algorithms—digital conductors that can orchestrate learning in both organoid and AI systems simultaneously. Brain organoids, in this grand plan, will need real-time feedback on their neural responses and performance, especially during complex reasoning and problem-solving tasks. AI systems in this setup can't just be answer machines; they need to be interpreters and teachers, explaining their thought processes in a language the organoid can understand.

Integrating organoids with electronics, data networks, AI algorithms, and robotics is a grand challenge. This cybernetic integration requires not just technical finesse but also conceptual advances in recognizing both living tissue and software as adaptable, learning entities.

Despite the extensive experimental optimization required and the risks involved, the seamless integration of biology and technology holds immense potential.

The Potential of OI

The emergence of OI represents an exciting development in the field of artificial intelligence, combining advancements in microphysiological systems (MPS) with computational approaches. MPS are laboratory models that aim to mimic the structure and function of human brain tissue on a small scale using organoids. These organoids contain multiple types of brain cells that interact with each other and are equipped with microfluidic components and sensors to maintain a more realistic environment and measure cellular activity.

The use of MPS in OI research allows scientists to create more representative models of human brain tissue, enabling them to study the interactions between brain cells and explore the potential of these neural networks for information processing.

This convergence of MPS and computational capabilities opens up new avenues for understanding brain functions, improving disease modeling, and potentially discovering new therapeutic interventions. As this technology matures, it may also contribute to the development of more advanced, adaptable, and efficient computing systems, with potential applications in various fields such as artificial intelligence and personalized medicine.

Significant research and technological advancements are needed before OI can fully realize its potential to revolutionize computing systems or have a transformative impact on our understanding of the human brain. Nonetheless, the combination of MPS and computational approaches in OI represents an exciting frontier in both neuroscience and artificial intelligence research.

The Miniature World of Organoid Engineering

In brain organoid engineering, scientists are still having challenges in growing human forebrain organoids complete with an assortment of brain regions and cell types.[15] One major hurdle is the lack of blood vessels, which limits the supply of oxygen and nutrients to the innermost regions of the organoids. To address this, researchers are experimenting with adding specialized cells that form blood vessels into the organoids. While this is a significant step forward, there's still a long way to go before lab-grown brain organoids can truly mirror the intricacies of the human brain.

When it comes to interfaces, high-density CMOS (complementary metal oxide semiconductor) nanowire arrays and optical techniques are the super-spies, letting us watch calcium dynamics dance across organoids with the precision of single-cell resolution. They provide the technological bridge that enable scientists to interact with, measure, and potentially influence these brain organoids at an unprecedented scale. Optogenetics and ultrasound, meanwhile, are like having a remote control for neurons, selectively stimulating brain cell activity.

Researchers have started playing "light and learn" with organoids by introducing light-sensitive proteins.[16] It's almost like teaching these brain organoids to respond to a game of cosmic peekaboo. By associating light cues with electrical stimuli, they're teaching organoids early pattern recognition, much like teaching a baby to recognize a favorite toy.

Merging organoid networks with AI represents the cutting edge of this field. AI software agents are learning to interpret organoid activity and respond adaptively, in a conversation between biology and technology. It's a partnership where each enhances the other, showcasing a level of complex pattern recognition that's like a duet between Mozart and a supercomputer.

The Astonishing Pace of Progress

OI is moving at an unexpectedly rapid pace, a technological sprint, highlighting the ambitious fusion of biology and technology. The goals set by researchers are ambitious stepping stones across a landscape filled with unknowns. Scaling up brain organoids to the complexity of the human brain, achieving real-time recording and stimulation of large cell groups within these organoids, and developing wireless, high-bandwidth interfaces are monumental tasks.

These endeavors aim to bridge the gap between the biological intricacies of the living brain and the computational prowess of mechanical minds. The vision extends to biocomputers that could rival human cognition, enabling direct brain-to-brain communication and creating autonomous robotic systems with adaptability equivalent to that of humans. This phase of research is not merely about technological innovation but about crafting a future where biology and technology merge seamlessly, challenging our current understanding of life and intelligence.

Concluding Thoughts

As we stand on the cusp of a new era, where the once-fictitious notion of merging organoid and AI technologies with our biological essence is becoming an increasingly tangible reality, we find ourselves faced with a future that is as exhilarating as it is uncertain. The implications of this fusion are nothing short of profound. The potential for enhancing human cognitive abilities and creating hybrid brains opens a world of possibilities that were once the stuff of science fiction.

In the next chapter, we will explore how brain organoids are already beginning to provide us with unprecedented insights into the intricacies of the human brain. These tiny, lab-grown structures are offering a glimpse into the fundamental mechanisms that underlie our thoughts, emotions, and very existence, shedding light on the complex processes of brain development and the origins of various brain disorders.

CHAPTER 3
NEURAL NURSERIES

As we delve deeper into the intricacies of brain organoids, we begin to unravel the mysteries of human cognition and pave the way for groundbreaking advancements in fields ranging from medicine to robotics. These tiny, three-dimensional structures, meticulously crafted to mimic the architecture of the human brain, hold the key to unlocking the secrets of our minds and ushering in a new era of intelligent systems.

At the heart of this endeavor lies the intersection of neuroscience and artificial intelligence, two fields that have long sought to understand and replicate the wonders of human cognition. Brain organoids, with their intricate networks of neurons and capacity for growth and adaptation, offer an unprecedented opportunity to bridge the gap between biological and artificial intelligence. By studying brain organoids, we can gain insights into the fundamental mechanisms of learning, memory, and cognition, and apply this knowledge to the development of more sophisticated and efficient AI systems.

However, the path to creating truly intelligent organoids is fraught with challenges and uncertainties. As we embark on this journey, we must confront the significant hurdles that lie ahead, from the technical difficulties of replicating the brain's complexity to the ethical implications of creating entities with human-like cognition. It is a journey that will test the limits of our scientific understanding and force us to grapple with profound questions about the nature of intelligence, consciousness, and the very essence of what it means to be human.

The Endgame: A Harmony of Mind and Machine

The ultimate goal of brain organoid research is to achieve a seamless integration of organic and electronic intelligence—a beautiful symphony of biology and technology that pushes the boundaries of what is possible. If successful, this could herald a new age of intelligent systems that possess the adaptability, efficiency, and creativity of the human brain, combined with the raw processing power and scalability of modern computing.

Imagine a future where intelligent prosthetics can intuitively respond to the user's thoughts, enabling individuals with disabilities to regain lost functionality and independence. Picture advanced robotics systems that can learn and adapt to new environments with the same ease and fluidity as a human child, opening up new frontiers in exploration, manufacturing, and service industries. And envision powerful computer systems that can process vast amounts of data with the pattern recognition and insight of the human brain, revolutionizing fields such as healthcare, finance, and scientific research.

These are just a few of the potential applications that could emerge from the successful development of intelligent organoids. By unlocking the secrets of the brain and harnessing its power in lab-grown structures, we could usher in a new era of technological advancement that transforms every aspect of our lives. It is a vision that inspires and motivates researchers around the world, driving them to push the boundaries of what is possible and to overcome the challenges that lie ahead.

The Art of Learning with Minimal Data

One of the most significant challenges in creating intelligent organoids is enabling them to learn and generalize from minimal data, much like humans do. Our brains are remarkably efficient at extracting meaningful information from limited experiences, allowing us to make inferences and predictions based on a small number of examples. When we encounter a new object or situation, we don't just perceive it as a static snapshot; instead, we experience a rich tapestry of sensory input that includes sight, sound, touch, and even smell. Our brains seamlessly integrate this multisensory information, creating a holistic representation that captures the essence of the experience.

This ability to learn from limited data is a hallmark of human intelligence, and it stands in stark contrast to the way most artificial neural networks operate. Traditional AI systems typically require vast amounts of labeled data to train on, often consisting of thousands or even millions of examples. While these systems can achieve impressive results in specific domains, they often struggle to generalize to new situations or to learn from a small number of examples.

Creating brain organoids that can learn from limited data poses a significant challenge, as it requires us to replicate the rich sensory experience and multisensory integration that characterizes human learning. Simply providing a small number of static images or data points may not be sufficient to capture the complexity and nuance of real-world experiences. Instead, we may need to provide organoids with more immersive and interactive environments that engage multiple sensory modalities and allow for active exploration and experimentation.

Towards a Multisensory Approach for Organoid Intelligence

This realization highlights the intricacies of biological learning and the limitations of current AI approaches. It suggests that creating truly intelligent brain organoids may require a fundamental shift in our understanding of learning and a move towards more embodied and experiential approaches. By studying how the brain learns from limited data and applying these insights to the development of organoid intelligence, we may be able to create systems that exhibit the same flexibility, adaptability, and efficiency as the human mind.

The journey towards assembling lab-made thought is still in its early stages, and the path ahead is long and uncertain. While significant progress has been made in recent years, with the development of more complex organoid structures and the integration of advanced technologies, there is still a vast distance between our current capabilities and the goal of creating truly intelligent organoids.

One of the major challenges that lies ahead is the sheer complexity of the human brain. With its billions of neurons and trillions of synaptic connections, the brain is a marvel of biological engineering that has evolved over millions of years. Replicating this complexity in lab-grown structures is a daunting task that will require significant advances in our understanding of neuroscience, as well as the development of new tools and techniques for manipulating and studying brain tissue.

Another challenge is the need for more sophisticated interfaces between brain organoids and the external world. To create brain organoids that can learn from experience and interact with their environment, we will need to develop advanced sensory and motor interfaces that can translate the language of neurons into the language of machines. This will require close collaboration between neuroscientists, engineers, and computer scientists, as well as the development of new technologies for sensing, stimulation, and feedback.

Despite these challenges, there are reasons for optimism and excitement about the future of organoid intelligence research. The rapid pace of progress in recent years, combined with the growing interest and investment in this field, suggests that we are on the cusp of a major breakthrough. As we continue to integrate discoveries from neuroscience, AI, and other fields, we may find new paths to overcome the obstacles that currently stand in our way.

Concluding Thoughts

As we ponder the future of brain organoid research and the challenges that lie ahead, we are naturally led to consider the role of AI in this endeavor. The intersection of neuroscience and AI holds immense potential for advancing our understanding of the brain and creating intelligent systems that blend the adaptability of biological cognition with the processing power of modern computing.

In the next chapter, we will explore more deeply the world of AI and its partnership with brain organoids. We will delve into the ways in which AI can serve as a collaborator and conductor, guiding the growth and development of organoids and extracting valuable insights from their complex neural activity.

This fusion of biology and technology promises to revolutionize our understanding of the brain and open up new frontiers in fields ranging from medicine to robotics. However, it also raises profound ethical questions about the nature of intelligence, consciousness, and the boundaries between the natural and the artificial.

CHAPTER 4

SYNAPSE AND SILICON: DECODING NEURAL NETWORKS

We now transition from the role of brain organoids as solitary detectives unraveling neurological enigmas to their partnership with a new ally—advanced computational techniques. Consider a world where innovative algorithms aren't just tools but collaborators, partnering with brain organoids to unlock mysteries of human cognition. These cutting-edge methods, known for their unparalleled pattern recognition capabilities, now venture beyond digital realms, interfacing with the biological intricacies of organoids. This fusion, reminiscent of a well-orchestrated symphony, promises groundbreaking insights into neural biology.

Picture these computational approaches as the seasoned conductor of this symphony, with organoids as the orchestra, each contributing unique notes to our understanding of the brain. This partnership goes beyond mere interaction; it's a harmonious blend of biology and technology, offering unprecedented perspectives on human development and disease.

In this world, brain organoids are more than just passive subjects; they're active participants. These brains organoids, thriving in specialized vessels, are not left to the whims of chance. Instead, they're nurtured in environments controlled by sophisticated algorithms—environments that are responsive, precise, and tailored to each organoid's needs. The computational techniques in this scenario act not just as passive observers but as active nurturers, constantly adjusting conditions to optimize organoid growth and development.

Yet, this journey isn't without challenges. It's a path laden with technical complexities and profound ethical questions. As we navigate this labyrinth, the excitement of discovery is tempered by the responsibility of innovation.

The Challenges of Managing Brain Organoids

Brain organoids in specialized vessels present a fascinating yet complex world, mirroring the intricate interactions found in a dynamic ecosystem. These miniature models of neural activity pose significant challenges in terms of management and cultivation. Just like balancing an ecosystem, controlling these organoids requires a nuanced understanding of the many factors that influence their development.

As we learned in the first chapter, the primary challenge lies in the precise orchestration of various biochemical and biophysical elements. These factors play a crucial role in shaping the growth and behavior of the organoids. Each organoid, with its unique cellular complexity, demands meticulous attention to ensure its healthy development, and the slightest imbalance can have significant consequences. This delicate task involves more than just monitoring; it requires a deep understanding of the interplay between numerous variables that drive organoid growth.

Technological Advancements in Organoid Management

To tackle these problems, researchers have made exciting progress by creating new tools and techniques. They've developed small-scale devices and artificial environments that provide just the right conditions for organoids to grow and thrive. One example is something called "Brain organoids-on-chip."[17] These combine 3D brain organoids with tiny, specialized devices that help create an environment that's much closer to what's found in a real brain, compared to traditional cell culture methods. The microfluidic devices recreate physiological conditions like fluid flow, oxygen and nutrient gradients, and removal of waste products, which are critical for proper organoid growth and survival. This leads to better growth conditions, more realistic results, and the ability to automate and speed up experiments, resulting in improved viability, reduced variability, and accelerated maturation of the organoids.

Additionally, brain organoids-on-chip are compatible with high-throughput screening and automation. The organoids can be cultured consistently at scale and analyzed using non-invasive techniques. This allows for more efficient and reproducible experiments to study brain development, disease modeling, and drug testing.

Scientists are starting to use AI to analyze huge amounts of data from brain organoid experiments.[18] This helps them figure out the best conditions for growth and could even help them grow new types of organoids. They're also using computer vision models—basically, teaching computers to "see" and understand images—to watch how brain organoids grow and change over time without disturbing them.

By combining AI with these new microfluidic technologies, researchers can process and make sense of complex biological information more easily. AI can spot patterns and trends that humans might miss, potentially leading to new discoveries about how organoids develop and function.

These cutting-edge approaches are revolutionizing organoid research, giving scientists unprecedented control over these tiny structures. As a result, we're moving closer to using organoids to study diseases, test treatments, and even develop personalized medicine tailored to individual patients. It's an exciting field with a lot of promise for improving human health in the future.

In the world of scientific exploration, where intricate neural networks take center stage, brain organoids emerge as vibrant ecosystems of neuronal activity. These miniaturized models of the human brain present a complex world to be decoded. Understanding the nuances of this neural activity is a formidable challenge.

In 2022, a groundbreaking study published in the scientific journal *Neuron* introduced the DishBrain system, a pioneering platform that allowed scientists to study how biological neural networks learn and adapt.[19] By placing these neural networks in a simulated game environment, like the classic video game "Pong," and analyzing their activity, DishBrain demonstrated that neurons can change and adapt their firing patterns based on the sensory information and feedback they receive. This exciting discovery revealed important clues about how biological intelligence works and how neural networks can display goal-oriented behavior.

The researchers carefully examined the neural activity data collected by the DishBrain system and gained valuable insights into the adaptive abilities of biological neural networks. The study showed that both human and mouse neurons can "learn" to play Pong by adjusting their activity patterns over time, with human neurons displaying better performance than mouse neurons. These findings highlighted the potential of the DishBrain system as a tool for exploring the differences in information processing abilities between different species.

The DishBrain system represented a novel approach to interfacing biological neural networks with artificial systems, potentially paving the way for future advancements in bridging the gap between artificial and biological intelligence. By providing a controlled setting for studying the behavior of biological neurons, this innovative platform opened up new possibilities for investigating the underlying mechanisms of learning, memory, and adaptation at the cellular level. The knowledge gained from DishBrain could potentially contribute to the development of more sophisticated artificial neural networks and enhance our understanding of brain function and disorders.

Following the success of the DishBrain system, Cortical Labs, the company behind the technology, recognized the need to further improve and optimize the platform. In practical terms, DishBrain was a groundbreaking but now redundant synthetic biological intelligence prototype. To better realize the potential of the technology and enhance accessibility, Cortical Labs developed a new device called the CL-1.

The CL-1 device is a significant advancement over the previous DishBrain system, offering a more sophisticated and accessible platform for studying biological neural networks. This innovative device combines a life-support system that keeps the brain cells alive with advanced technology that can both stimulate the cells and record their activity in real time. Compared to the DishBrain system, the CL-1 can connect to many more brain cells simultaneously and can send and receive signals much faster, allowing for more complex experiments and more precise data collection.

Accelerating Discovery Cycles

We are entering an age in which the pace of discovery will dwarf that of the Renaissance, transforming the methodical progression of scientific inquiry into a rapid-fire succession of breakthroughs. This acceleration is spearheaded by the emergence of closed-loop discovery algorithms, the new powerhouses in the world of research.[20] Imagine a smart robot scientist that can come up with its own ideas, do experiments to test them, and then learn from the results to come up with even better ideas. It does this over and over, getting smarter and making discoveries faster than any human scientist could. Closed-loop discovery algorithms use AI to automate the entire research process, from hypothesis to experiment to conclusion, in a continuous loop. This allows them to make scientific breakthroughs at a pace that was once unimaginable.

In OI, closed-loop algorithms could be the key to unlocking the secrets of the brain. By applying these powerful computational tools to the analysis of brain organoids, researchers could gain unprecedented insights into the complex workings of these brain organoids. Imagine a system where AI-guided experiments continuously generate data from brain organoids, and the results are fed back into the AI models, refining them and leading to more accurate predictions about neural pathways and brain functions.

This closed-loop approach to studying brain organoids could accelerate our understanding of how the brain develops, learns, and adapts. By automating the experimental process and using AI to analyze the vast amounts of data generated by these organoids, researchers could identify patterns and uncover fundamental insights into brain function at a pace that was once unimaginable. This could lead to breakthroughs in fields ranging from neuroscience to medicine, paving the way for new treatments for neurological disorders and a deeper understanding of human cognition.

Moreover, closed-loop algorithms could be used to optimize the integration of brain organoids with digital components. This could lead to the development of more sophisticated hybrid systems that combine the strengths of biological and artificial intelligence. These systems could potentially exhibit human-like cognition and problem-solving abilities, opening up new frontiers in fields such as robotics and artificial intelligence.

The Irreplaceable Value of Human Insight

However, amidst this whirlwind of innovation, it's crucial to remember the irreplaceable value of human insight. While automated systems excel in efficiency and speed, they cannot fully replicate the depth and nuance of human creativity and intuition. Even if a robot with a brain organoid could understand complex topics and conduct research semi-autonomously, it would likely still lack the rich inner world, self-awareness, and ability to make intuitive leaps that characterize human cognition. These tools, as advanced as they are, serve as extensions of our own intellectual and creative capacities.

They may be able to mimic certain aspects of human thought to an impressive degree, but there remain (at least for the foreseeable future) important qualitative differences between organoid-based AI and the human mind that conceived it. No matter how sophisticated they get, they still fundamentally depend on and exist to serve human ingenuity rather than to supplant it entirely.

Integrating OI into this technological framework adds a vital dimension of biological authenticity. This is crucial because it allows AI systems to analyze and learn from data that more closely represents actual human brain activity and development. This authenticity enhances the relevance and potential applicability of AI-driven insights to human neuroscience and medicine. These brain organoids offer a window into the complexities of neural pathways and brain functions. AI, in analyzing this data, extends our understanding to previously unreachable depths.

In this symbiotic relationship, humans are far from being mere onlookers. We are the directors, guiding AI to interpret the subtle intricacies of brain activity as reflected in the brain organoids. This partnership doesn't just enhance our understanding of neural functions and disease mechanisms; it propels us towards a deeper comprehension of consciousness itself.

Thus, in this era of rapid discovery, we find a harmonious blend of technology and human intellect, each complementing the other to explore the deep mysteries of the brain.

Quantum Computing: A New Frontier in Decoding Neural Networks

As we ponder the challenges of decoding the intricate world of brain organoids, another emerging field offers an exciting glimmer of hope:

quantum computing. This revolutionary technology harnesses the strange properties of quantum mechanics to process information in fundamentally different ways from classical computers. By leveraging principles like superposition and entanglement, quantum computers have the potential to solve certain problems exponentially faster than even the most powerful supercomputers today.

Superposition, a core concept in quantum computing, refers to the ability of a quantum system to exist in multiple states simultaneously. While a classical computer bit is either a 0 or a 1, a quantum bit (qubit) can be both 0 and 1 at the same time. This allows quantum computers to perform many calculations in parallel, vastly increasing their processing power.

Entanglement, another key quantum principle, occurs when two or more particles become inextricably linked, such that the state of one particle cannot be described independently of the others. This means that measuring one entangled particle instantly affects its counterpart, regardless of the distance between them. Entanglement enables quantum computers to perform certain computations that would be impossible for classical computers.

Quantum Computing's Potential in Brain Organoid Research

The immense computational power offered by quantum computing could be a game-changer in analyzing the vast amounts of complex data generated by brain organoids and their interactions. Quantum algorithms, designed to exploit superposition and entanglement, could potentially uncover hidden patterns, correlations, and insights that would be intractable for classical computing methods.

Moreover, the inherent complexity of quantum systems may make them particularly well-suited for modeling and simulating the intricate dynamics of biological neural networks. By harnessing the power of quantum computing, researchers could potentially develop more accurate and efficient models of brain function, leading to breakthroughs in our understanding of cognition, disease, and drug discovery.

As the fields of brain organoids, artificial intelligence, and quantum computing continue to advance and intersect, we may be on the cusp of a new era in neuroscience. The convergence of these cutting-edge technologies holds the promise of unraveling the mysteries of the human brain and ushering in a future of unprecedented medical and scientific progress.

Applications of Quantum Computing in Brain Organoid Research

One possible way quantum computing could help is by using specialized algorithms to analyze the enormous amounts of data produced by large-scale brain organoid experiments. These algorithms could quickly identify subtle patterns and connections that might be overlooked by conventional computing methods. For instance, quantum machine learning algorithms could be used to study gene expression data from brain organoids at different stages of development. By processing this information using a quantum approach, researchers might discover hidden relationships between gene networks and the formation of specific types of brain cells or circuits. This could offer valuable insights into the complex genetic and environmental factors that influence brain development.

Another exciting possibility is the use of quantum computers to simulate the behavior of brain organoids themselves. By creating quantum models that capture the intricate dynamics of neural networks, researchers could virtually test hypotheses and predict the effects of various interventions on brain function. These simulations could be far more efficient and accurate than those performed on classical computers, as quantum systems are inherently well-suited to modeling the complex, probabilistic nature of biological systems.

Quantum-enabled brain organoid simulations could be used to study the impact of drugs, toxins, or genetic mutations on neural development and function. By observing how these virtual organoids respond to different perturbations, researchers could gain valuable insights into the mechanisms underlying various neurological disorders and identify potential therapeutic targets. This could accelerate the drug discovery process and lead to the development of more effective, personalized treatments for brain diseases.

Furthermore, the integration of quantum sensors with brain organoid platforms could enable the real-time monitoring of neural activity at an unprecedented level of sensitivity. Quantum sensors, such as nitrogen-vacancy centers in diamond, can detect minute changes in magnetic fields generated by neural firing. By embedding these sensors into brain organoids, researchers could map the formation and evolution of neural circuits with exquisite spatial and temporal resolution. This could provide new insights into how information is processed and stored in the brain, and how these processes might be altered in disease states.

As the fields of brain organoids, artificial intelligence, and quantum computing continue to advance and intersect, we may be on the cusp of a new era in neuroscience. The convergence of these cutting-edge technologies holds the promise of unraveling the mysteries of the human brain and ushering in a future of unprecedented medical and scientific progress.

However, as we explore these cutting-edge technologies and their applications in neuroscience, it's crucial to maintain a balanced perspective. While it's easy to get caught up in the hype surrounding these developments, we must also be mindful of the current limitations and challenges that lie ahead. It's essential to keep in mind that this integration of quantum computing and brain organoid research is still mostly theoretical and in the very early stages of development. While quantum computing holds tremendous potential, the technology is still young, and significant progress is needed before it can be realistically applied to studying brain organoids.

Avoiding Hype and Capability Overreach

In addressing the buzz around advanced biocomputational techniques and organoids, it's essential to differentiate between what's currently achievable and what remains firmly in the realm of science fiction. The notion that today's basic organoids and AI might suddenly give rise to a human-like consciousness is more fantasy than fact. It's similar to expecting a beginner's simple melody to rival a symphonic masterpiece. This perspective overlooks the significant scientific challenges and the vast understanding still needed to grasp subjective experience and advanced cognition.

While it's not impossible that future advancements could lead to groundbreaking developments, our immediate focus should be on leveraging these technologies for practical and achievable goals, such as drug discovery and disease modeling. This approach is like laying the foundation stones of a building, ensuring a solid base before aspiring to reach the skies.

While it's exciting to speculate about future possibilities, we must ground our expectations in the reality of current capabilities. Emphasizing practical applications over sensationalist predictions will ensure that our journey in AI and organoid research is marked by steady and meaningful progress, rather than premature, risky leaps into the unknown.

Concluding Thoughts

Advanced computation techniques and brain organoids herald a new dawn in neuroscience. Our journey through this emerging landscape has revealed the transformative potential of these synergies of silicon and synapse not just as tools but as collaborators in organoid research, offering profound insights into the complexities of the human mind.

This exploration has illuminated the symbiotic relationship between advanced computational techniques and organoids, highlighting how their integration can enhance our understanding of neural networks and disease mechanisms. AI, acting as an insightful conductor, has shown its capability to interpret and respond to the nuanced biology of organoids, fostering a harmonious fusion of technology and biology. Yet, this journey also brings to light the challenges ahead, including the need for precision in organoid management, ethical considerations in merging biology with AI, and the balance between ambitious speculation and the realities of current technology.

We look ahead with anticipation to the next phase in our quest to decode the human brain. The integration of advance biocomputational techniques and organoids has laid the groundwork for breakthroughs that once seemed unattainable, blurring the lines between the biological and the technological and paving the way for novel paradigms of knowledge.

PART II

OI: REVOLUTIONIZING LIFE AS WE KNOW IT

CHAPTER 5

INSIGHTS FROM LAB-GROWN MINDS

In this chapter we will explore how brain organoids can help us unravel the mysteries of the most intricate and complex organ in the human body—the brain. These tiny, lab-grown marvels are beginning to provide us with revealing insights into the intricate architecture of a real brain, shedding light on the fundamental mechanisms that underlie our thoughts, emotions, and very existence. As we delve deeper into the world of brain organoids, we will discover how these miniature models are revolutionizing our understanding of brain development and various brain disorders, offering new avenues for studying the causes and potential treatments for these conditions.

Replicating Brain Complexity on a Small Scale

Previously, we saw that scientists have been busy in their laboratories, crafting something that's revolutionizing our understanding of neuroscience: brain organoids about the size of breadcrumbs, which are filled with immense potential. Under the microscope, it's a full-scale spectacle, showcasing the cellular ballet of new nerve cells, or neurons, in an organoid. These brain organoids are creating early brain structures that mirror the brain's own intricate form, offering a novel approach to understanding developmental brain disorders.

Brain organoids are not trying to replicate an entire brain. Instead, their role is to reconstruct specific aspects of brain biology, providing a new model for exploring the complex dance of mind development. They are a valuable tool for studying human neurobiology, bridging the gap between in vitro and in vivo research approaches.

These organoids offer just enough biology for scientists to probe the mysteries of the brain's construction and the occasional missteps that can occur in it. They allow us to peek into the once impenetrable realm of the brain, growing tailored samples that reveal previously hidden insights.

For example, researchers are exploring how the brain develops its convoluted folds.[21] These folds enable higher cognition by increasing the brain's surface area within the limited space of our skulls. By modeling this wrinkling process in organoids, we are beginning to understand the forces that sculpt the fetal brain's smooth surface into a corrugated operations center of thought.[22] This folding is essential for higher thinking and complex functions. By studying the growth of brain cells, the molecular processes that control folding, and the genetic factors influencing brain size and patterns, scientists can gain insights into fundamental brain formation.

This research has significant implications for understanding and treating brain disorders linked to folding issues, as well as conditions like schizophrenia and autism. By creating models that mimic brain folding, scientists explore how disruptions may lead to these disorders and develop potential treatments. Additionally, insights into brain folding improve imaging techniques, leading to better brain maps and more accurate diagnoses. Understanding the relationship between brain structure and function could connect specific folding patterns to abilities or behaviors. Ultimately, this work paves the way for new therapies targeting genes or pathways involved in brain folding, offering hope for advancements in brain health and our understanding of human cognition.

Additionally, brain organoids are shedding light on the step-by-step process behind the brain's vertical layering of specialized neurons. It's like deciphering the brain's own Rosetta Stone, uncovering how cells migrate and form distinct circuits during development, crucial for functions ranging from movement to reasoning. By studying brain organoids, researchers are gaining unprecedented insights into how neurons migrate and form the brain's layered structure, which not only advances our understanding of neurodevelopmental disorders but also improves drug testing, paves the way for personalized treatments, and potentially informs advancements in AI mimicking neural circuit formation

By making visible the transformation of stem cells into brain organoids, the once-hidden forces shaping our minds, identities, and potential are coming into clearer view. One could compare it to having a portal into the genesis of our being, shedding light on the bottom-up construction of our brains. In this groundbreaking endeavor, scientists are doing more than creating a model of the brain; they're unraveling the recipe of our very essence.

Understanding the Roots of Brain Disorders

Brain organoids are starting to take center stage in the field of neuroscience. They offer us unprecedented insights into the complex worlds of autism, schizophrenia, dementia, and other mental illnesses. These conditions, deeply entrenched in the complex neural networks of the brain, have been the focus of intensive study. Now, with the advent of brain organoids, our understanding is not just inching forward; it's taking leaps and bounds.

Historically, scientists have been like detectives in a cosmic mystery, attempting to unravel the causes of these mental illnesses primarily through genetics and postmortem studies. However, the introduction of patient-derived organoids has revolutionized this process. For the first time, researchers can directly observe the onset and progression of mental illnesses in living tissue, tracing how they begin to alter neural pathways. This approach provides a new depth of understanding, offering a significant stride towards developing effective therapies for psychiatric conditions.

Concluding Thoughts

Brain organoids are playing a key role in unraveling the mysteries of the human brain. It's clear that these tiny marvels hold immense potential not just for basic research but for transforming medicine as we know it.

By providing unprecedented insights into brain development and the origins of neurological disorders, brain organoids are already beginning to reshape our understanding of the most complex organ in the human body. But this is just the beginning. As we'll explore in the next chapter, these lab-grown wonders are poised to revolutionize the way we approach everything from drug discovery to personalized therapies.

Imagine a future where we can test the safety and efficacy of new treatments on miniature models of the human brain, without putting patients at risk. Or where we can study the effects of toxic exposures on fetal brain development and develop strategies to prevent birth defects. With brain organoids derived from a patient's own cells, we may even be able to create tailor-made treatments for a wide range of neurological conditions.

But as we'll see, realizing this vision of a medical revolution won't be without its challenges. Fully harnessing the lifesaving potential of brain organoids will require not just scientific ingenuity, but a willingness to grapple with profound philosophical and ethical questions. What moral status should we accord to these lab-grown brains? How do we balance the imperative to innovate with the need to proceed cautiously and responsibly?

As we stand on the cusp of a new era in medicine, these are the questions we must confront. The study of brain organoids has already opened up a world of possibilities, and the discoveries that await us in the coming years have the potential to transform not just our understanding of the brain, but the very way we approach healing.

In the next chapter, we'll start to map this new terrain, exploring the myriad ways in which brain organoids are set to reshape medicine, and the challenges and opportunities that lie ahead.

CHAPTER 6

MEDICAL REVOLUTION: DISH-DERIVED HEALING

For over a century we've rallied against illness using the same imperfect weapons: questionable animal experiments and risky trial-and-error medicine unleashed upon patients. Hardly precise strikes, no? But now reinforcements have arrived! For brewing within humble specialized vessels lies ammunition for a bona fide medical revolution! As we have learned in earlier chapters, these brain organoids are remarkable miniscule models of human organs ingeniously crafted from ordinary stem cells. Much like microchips radically shrunk computers, these living micro-organs apply miniaturization to medicine and stand poised to disrupt everything from drug discovery to curative therapies.

Just imagine identifying promising pills years faster without endangering patients. Or averting birth defects by revealing toxic exposures early. Tailor-made disease treatments could even emerge from organoids derived from your own cells. The possibilities are endless!

But revolutions seldom unfold easily. Fully tapping brain organoids' lifesaving potential will require radically rethinking both healthcare and our very definitions of life itself. What responsibilities do we have to these cutting-edge lab-grown organs? Do emerging brain organoids deserve cautious protections? Could they someday feel or think themselves? Such philosophical brain-twisters multiply by the minute!

Hacking the Limits of Animal Models

Here's pharma's dirty secret: those popular drug ads depicting lab coats and glossy mice hide a clumsy reality: animal trials frequently fail to predict a drug's effects in people. Genetic differences mean mouse models simply don't capture nuanced human biology. No wonder approval rates for new drugs hover at less than 12%, after billions are wasted on candidates that flop in human tests.[23]

Emerging brain organoid technologies allow us to model diseases and develop drugs in miniaturized human tissues instead of imperfect animal substitutes. Grown from stem cells, organoids recreate intricate structures and function amazingly like our organs in vitro, enabling unprecedented insight into human biology.

Companies are harnessing brain organoids to potentially screen drug candidates at high throughput, rapidly weeding out duds while advancing the most promising.[24] This organoid filtering helps de-risk pipelines and boost success rates for human trials. With further refinement, organoid testing may one day fully replace animal models, bringing faster cures without sacrificing human relevance.

Beyond improving drug pipelines generally, brain organoids also open the door to therapies tailored to match each patient's unique genetics, an approach called precision medicine.[25] Picture brain organoids derived from your own cells that sensitively model your distinctive biology before you ever pop a pill. Such "clinical trials in a dish" could determine your ideal medications, doses, and combinations. Suddenly, the era of crude one-size-fits-all medicine looks primitive.

Living up to brain organoids' full disruptive potential will require confronting hurdles head-on. Manufacturing processes will need improving to provide scalable, reproducible brain organoids that reliably mimic human tissues. Genetic and microenvironmental cues guide differentiation, but better control is still needed. It will require thoughtful policies for transitioning from animal models. Significant challenges will remain, but the lure of brain organoid technologies promises to reduce drug development windows from decades to years, which will significantly reduce costs. Not long from now, your life-saving medicine's origin story may trace back to pioneering trials in lab-grown miniaturized brain organoids.

Brain Organoids to the Rescue: Tackling Disease Through Miniature Models

Beyond improving drug development, brain organoids provide us with an unparalleled view into human disorders, which can help tackle conditions like Alzheimer's and autism at their biological source. These miniaturized models make it possible to track pathology step-by-step in human tissue. Early applications are already unraveling mysteries on which progress has been stalled for decades.

Brain organoids used in researching Alzheimer's have displayed the intricate tango between neuron-killing amyloid plaques and tangles in the development of the condition, and which drugs must therefore target in tandem.[26] This improved model explains why compounds only addressing plaques failed again and again in clinical trials. No wonder a cure has seemed so elusive! Brain organoids are now yielding pivotal insights to get treatment back on track.

Autism organoids are also shedding light on risks and biological pathways.[27] By comparing organoids formed from stem cells of healthy individuals with those from autistic individuals, researchers are gaining precious clues into how wiring and function go awry in developing brains, converting ambiguity into defined mechanisms. This understanding promises to reveal subgroups and suggest biological targets for therapy. We will delve deeper into these groundbreaking advancements and their implications for the future of neuroscience in the next chapter.

Brain organoids should complement clinical research, not replace it. Each approach has limitations. However, used together, they promise to unravel more disease mysteries in the coming decades than all prior centuries combined. The era of groping blindly without maps or models is coming to an end, as organoids can offer hope of tackling everything from Alzheimer's to autism by revealing their biological intricacies in human tissues

Concluding Thoughts

In this chapter, we've revealed organoids' potential to revolutionize the landscape of healthcare and medical research.

Brain organoids herald a significant shift from traditional, often imprecise medical methods. Their role in drug discovery and the development of personalized medicine represents not just a technological advancement but a paradigm shift, redefining our approach to health and disease. From rapidly identifying effective medications to tailoring treatments based on individual genetic makeup, organoids are at the forefront of a medical renaissance.

We will now explore how organoids are shedding new light on some of the most perplexing and elusive brain disorders, and how these neurons in a dish translate into tangible advances in our quest to conquer the brain's most elusive disorders.

CHAPTER 7

NEUROLOGICAL CLARITY: ROLE OF LAB NEURONS

Our focus now zeroes in on the neurodegenerative and psychiatric landscapes. Here, brain organoids emerge as miniature detectives, unearthing the elusive culprits behind conditions like Alzheimer's and schizophrenia. This chapter delves into the role of lab-grown neurons in demystifying the brain's most perplexing disorders, offering a beacon of hope for millions affected by these conditions.

Watching a mind fade away is like witnessing a library burning, one precious volume at a time: this sparkling tapestry of thoughts and memories slowly vanishing in smoke. It's heart-wrenching. But imagine if we had a microscopic model of this library. That's what brain organoids offer: a chance to model the chaos and find ways to extinguish the fire.

Of course, expecting these brain organoids to fully mimic the intricacies of a human mind is a big ask. But in their own small yet significant way, they are cracking codes of diseases that were once considered indecipherable enigmas.

Personalized Organoid Training

Crafting a personalized brain organoid that replicates each individual's unique neural wiring is as daunting as knitting a custom sweater for every penguin in Antarctica. Yet, even without perfect personalization, these brain organoid models serve as powerful tools to unravel the mysteries of neurodegenerative and psychiatric disorders. Imagine working with organoids that possess limited cognitive and emotional capacities—much like simplified versions of the human brain. How might this simplicity influence their ability to model complex conditions like Alzheimer's or schizophrenia? Could cultivating organoids at various developmental stages, reminiscent of students progressing through school, enhance our understanding of disease progression? Perhaps by establishing specialized training protocols or research environments, we can amplify their capabilities. Exploring these possibilities deepens our journey into OI, bringing us closer to extinguishing the fires that consume the mind's precious library.

Illuminating the Darkness of Alzheimer's

In the grand tapestry of medical mysteries, Alzheimer's disease stands out like an enigmatic villain in a Sherlock Holmes story. It lurks in the shadows of our neurons, pilfering memories, personalities, and the very essence of self. But now, the humble brain organoid, a miniature superhero no bigger than a breadcrumb, is bringing its flashlight to the murky corners of Alzheimer's.

In one remarkable study, researchers grew brain organoids from Alzheimer's patients with inherited genetic mutations.[28] Astonishingly, these tiny lab-grown brains went on to develop the defining features of Alzheimer's—sticky clumps of amyloid protein known as plaques and twisted strands of another protein called tau—in a pattern that faithfully recapitulated the sequence and timing of the disease's progression in human brains. Watching the organoids mature was like peering into a crystal ball to witness the future unfolding of Alzheimer's in a person's brain, a groundbreaking advance that opens up entirely new possibilities for unraveling how the disease takes root and finding ways to stop it in its tracks.

Other researchers have taken it a step further, using CRISPR gene editing to introduce Alzheimer's risk genes like APOE4 into organoids.[29] They found that this genetic risk factor disrupted the brain's immune cells (microglia) and support cells (astrocytes), impairing their ability to clear toxic amyloid. It's as if APOE4 ties the hands of the brain's clean-up crew.

But the revelations don't stop there. Other researchers showed that neurons in organoids derived from Alzheimer's patients were hyperexcitable, firing in a frenetic, uncoordinated way.[30] This neural pandemonium could help explain the cognitive dysfunction in Alzheimer's.

These tiny tissue wonders are not just revealing how Alzheimer's dismantles the brain, they're also serving as personalized test platforms for potential treatments. One study screened over 1200 existing drugs on Alzheimer's organoids, identifying a few that could rescue dying neurons.[31] It's like rapidly auditioning thousands of potential heroes to save the day.

As brain organoids become more sophisticated, incorporating blood vessels and immune cells, they'll paint an even more vivid and complete picture of Alzheimer's. In the future, doctors might use a patient's own cells to create personalized brain organoids and test tailored therapies, like having a stunt double for your brain.

Of course, the battle against Alzheimer's isn't just waged under the microscope. It's also fought in hospital wards and living rooms, with the disease exacting a heavy financial and emotional toll. A recent study estimated that the estimated healthcare costs associated with Alzheimer's disease treatment were $321 billion, with costs projected to exceed $1 trillion by 2050. This burden underscores the urgent need for the kinds of breakthroughs that brain organoids might one day deliver.[32]

Brain organoids are a key to unlocking Alzheimer's secrets. And while we haven't yet cracked the code, these tiny tissue wonders are bringing us closer than ever to understanding this relentless riddle of a disease and, hopefully, to writing a happier ending to its story.

New Light on Schizophrenia's Dissonant Chords

In the complex symphony of the mind, schizophrenia is like a mischievous interloper, introducing discordant elements that disrupt life's rhythm just as it's beginning to play. Enter again our tiny but mighty brain organoids, akin to a team of musical detectives, delving into the enigma of schizophrenia. These organoids are miniature replicas of neural tissues, derived from the very brains grappling with this disorder. In the hands of researchers, they become like crystal balls, revealing the chaotic dance of neurons that underpins schizophrenia's symptoms.

Recent advances in brain organoid research have shed light on the intricate development of schizophrenia at the cellular level.[33] Patient-derived brain organoids exhibit distinctive patterns of neural misfiring due to their ability to recapitulate the cellular and molecular dysfunctions characteristic of neurological disorders such as schizophrenia. For instance, a study involving organoids from schizophrenia patients demonstrated abnormal gene expression in pathways related to brain development, immune response, and energy production.[34] These organoids also showed a diminished response to electrical stimulation, mirroring the imbalance between excitatory and inhibitory signals seen in schizophrenia.

Furthermore, research using brain organoids from schizophrenia patients revealed early cortical maldevelopment, characterized by the abnormal scattering of developing neural cells. This dispersion was accompanied by a blockade of cortical neuron development, highlighting the neurodevelopmental dysfunctions in schizophrenia.[35]

It's like if the pianist's sheet music has been scrambled, notes haphazardly strewn about, and then they try to accompany an orchestra where every instrument is playing a different tune. The result? A bewildering symphony of the mind that we know as schizophrenia.

But here's where the plot thickens, and hope begins to shine. The use of brain organoids in studying schizophrenia and other neurological disorders offers significant potential for understanding the underlying molecular and cellular mechanisms. By analyzing these organoids, scientists can identify specific molecular culprits that contribute to the disorder, providing a basis for developing targeted therapeutic interventions.

Single-cell sequencing of schizophrenia organoids has identified cell-specific neuropathology, suggesting that schizophrenia's molecular underpinnings can be encoded on a cell-type-by-cell-type basis.[36] Moreover, proteomic analysis of schizophrenia-sourced organoids showed similarities to postmortem brain tissues from schizophrenia patients, further validating the use of organoids to replicate the molecular mechanisms of schizophrenia.[37]

Excitingly, brain organoids derived from schizophrenia patients have been used to explore the effects of new drugs and to verify the efficacy of treatments in a personalized manner.[38] This approach allows for the screening of drugs on patient-specific organoids, which can help in developing more effective and tailored therapeutic strategies.

These discoveries offer a chance to pharmacologically restore the brain's harmonious melody, much like retuning a piano, by addressing the chaotic neural pathways and turning the cacophony back into coherent music. This research underscores how organoids can be used as powerful models to unravel the complex molecular and cellular dysfunctions characteristic of schizophrenia, offering a new lens through which to view this perplexing disorder.

Organoid Insights into Autism's Divergent Minds

Brain organoids can also be grown from the cells of patients with autism spectrum disorder (ASD). They offer a unique window into the complex world of the autistic mind, revealing how brain connections develop differently in individuals with ASD.

Recent studies using brain organoids have shed light on the specific changes that occur in the brains of people with autism. For example, researchers have discovered an imbalance in the types of neurons that form during early brain development in ASD patients.[39] This imbalance may contribute to the differences in brain size and function often observed in individuals with autism.

Scientists have also used advanced gene-editing techniques, such as CRISPR, in combination with brain organoids to identify developmental defects associated with autism.[40] By studying these defects at a cellular level, researchers can pinpoint the specific types of brain cells that are most vulnerable in ASD.

Furthermore, brain organoids are being used to model specific regions of the brain that are known to be affected in autism, such as the forebrain.[41] By examining these regions in detail, scientists can better understand the molecular, cellular, and functional changes that contribute to the diverse symptoms of ASD.

So, how can this information be used to help individuals with autism and their families? First and foremost, by understanding the developmental mechanisms underlying ASD, researchers can develop targeted therapies that address the specific cellular and molecular dysfunctions associated with the disorder. Brain organoids also provide a platform for drug screening, allowing scientists to test the effectiveness of potential treatments in a human-specific context.

Moreover, because brain organoids can be grown from the cells of individual patients, they offer the potential for personalized medicine. This means that treatments could be tailored to the unique genetic and cellular profile of each person with autism, potentially leading to more effective interventions.

Beyond the medical applications, brain organoid research also has important ethical and social implications. By providing tangible, observable evidence of the developmental differences in the autistic brain, organoids help to shift the conversation about autism from abstract ideologies to a more factual, nuanced understanding of neurodiversity. This can promote greater acceptance and support for individuals with ASD and their families.

Elucidating Depression's Submerged Shadows

In the complex world of cognitive conditions, depression looms large, a shadowy figure capable of plunging even the most luminous minds into darkness. It's like an uninvited guest at life's banquet, turning every succulent course into bland fare. While brain organoids can't quite step into the shoes of subjective moods, they do offer intriguing windows into the biological underpinnings of our emotional lives. We can imagine these organoids as tiny submarines, diving into the murky depths of our neural seas. They're exploring the intricate ballet of chemicals that choreograph our emotions and motivations.

Scientists, acting as neuro-chemists, tinker with neurotransmitters—serotonin, the regulator of our inner peace treaty; dopamine, the conductor of our brain's reward symphony—teasing apart their roles in the grand orchestra of the mind. These insights gleaned from organoids help us piece together the picture of how chemical imbalances may cast long shadows over our emotional well-being. Armed with this knowledge, the pursuit of antidepressant therapies becomes less like casting a wide net in the hope of catching something and more like precision surgery. The goal? To tailor treatments that can recalibrate the delicate balance of neurotransmitters, restoring emotional equilibrium. But that's not all.

Brain organoids also serve as promising testbeds, allowing us to explore how external factors—medications, toxins, stress hormones—might remodel the mood-related landscapes of our neural tissue.[42] While this application is still in its early stages, it offers a potentially safer, more ethical, and more effective alternative to traditional animal models for such studies. As research progresses, we can envision exposing organoids to various external influences and observing the resulting changes in neurotransmitter systems, neural connectivity, and gene expression. These insights could illuminate the biological mechanisms of depression with unprecedented clarity, informing the development of more targeted, personalized interventions. However, it's important to note that comprehensive findings directly linking organoid studies to mood disorders are still limited. More empirical research is needed to fully substantiate these potential applications and translate them into tangible clinical advances.

Moving beyond traditional animal models, which often miss the mark in predicting antidepressant effects in humans, organoids stand as superior stand-ins for human brain chemistry. And what if these tiny neural explorers could unlock the secrets of emotional resilience? We could potentially craft interventions to fortify young minds against the storm clouds of future depressive disorders, building up a reserve of affective strength, cognitive flexibility, and neurochemical balance. Of course, it's important not to be naïve.

The organoid approach isn't a silver bullet. The tangled web of depression's causes—from societal pressures to existential angst—can't be untangled by science alone. Nonetheless, the promise of brain organoids as powerful platforms for probing the complex interplay between external factors, brain chemistry, and mood is increasingly apparent, heralding a future of more precise and effective depression treatments. As organoids illuminate the biological aspects of this mental morass, they will enable us to chart new courses for both biomedical and psychosocial exploration.

Illuminating Anxiety's Obscured Roots

In the sprawling landscape of mental health, anxiety disorders stand like a dense, uncharted forest, a realm where fear and worry grow wild and untamed. Unlike depression, with its well-trodden paths of biological investigation, the neural roots of anxiety are shrouded in a thicker fog.

Brain organoids can offer researchers a map and compass to explore this terra incognita. Think of these organoids as miniaturized recon drones, flying over the complex terrain of neurotransmitter systems, network connectivity, and developmental pathways that mirror the clinical manifestations of anxiety.

The idea of tweaking these organoids to mimic the absence of neuron receptors often implicated in anxiety is compelling. It would allow scientists to play a high-stakes game of Jenga, removing one block at a time to see how the tower reacts. Without the ethical and practical complications of human experimentation, these organoid models could become invaluable in understanding how cellular stressors might tilt the balance towards heightened vulnerability or resilience.

As the field progresses, this journey through organoid models could potentially unearth new pharmaceutical El Dorados, ripe with potential for targeting anxiety at its very roots. Time-lapse imaging of these organoids could be like fast-forwarding through a movie of neural development, revealing the pivotal scenes where neurodevelopmental deficits set the stage for hyper-reactive fear circuits.

Moreover, the malleability of stem cell engineering allows for the possibility of tailoring organoid models to specific subtypes of anxiety disorders in the future. While we're not yet at the point of turning these insights into concrete medical miracles, the journey of brain organoids is charting a course towards transformative discoveries. Each revelation, each glimpse into the elusive origins of anxiety, or identification of a neural Achilles' heel, brings us a step closer to a world where more individuals can navigate their way out of the thicket of fear and into the open spaces of equilibrium and peace. However, it is crucial to balance the excitement about the potential of brain organoids with a clear understanding of the current state of the research, acknowledging both the promise and the limitations of this approach. This approach is still largely theoretical and has not yet been widely implemented in research.

Meditating on the Mind–Body Connection

But the story of organoids doesn't end with the brain. They also serve as gateways into the enigmatic mind–body relationship, a domain ripe with implications for everything from chronic pain to psychogenic seizures.

Picture brain organoids not in isolation, but conjoined with other tissue types—muscle, vascular, or even intestinal models. These combinations are like miniature ecosystems, mirroring the intricate dialogues between our mental states and physical symptoms.

One fascinating example is the development of organoids to study the interaction between the gut and the brain.[43] These organoids can be tweaked to simulate stress responses, allowing researchers to observe how emotional states can manifest as gastrointestinal issues, and how gut health can affect mental well-being. This research could lead to novel treatments targeting the gut-brain axis for conditions like irritable bowel syndrome and anxiety.

The microbiota-gut-brain axis is a bidirectional communication system that involves neural, hormonal, and immunological pathways. Recent studies suggest that the gut microbiome, comprising a vast number of microorganisms, is essential for the development and maturation of brain systems associated with stress responses.[44] Variations in the microbiome, microbial substrates, and exogenous prebiotics, antibiotics, and probiotics have been linked to diverse neuropsychiatric disorders such as autism, depression, anxiety, and schizophrenia.[45]

Traditionally, animal models, particularly gnotobiotic mice, have been used to explore the mechanisms underlying the microbiome-gut-brain axis cross talk. However, ethical considerations and interspecies differences that limit the translatability of data to human systems have pushed researchers to seek alternatives. Advances in 3D cell biology, materials science, and bioengineering have led to the development of in vitro models that better mimic native tissue structure and function.

It's like watching a conversation between the brain and the body, understanding how a whisper in the mind can become a shout in the stomach. Such exploration might lead us to new pharmaceutical targets or innovative devices designed to harmonize the mind–body symphony. Imagine a future where the treatment of disorders is not just about addressing the physical or the psychological, but about tuning the intricate interplay between the two.

Of course, the riddle of somatization is complex, a puzzle unlikely to be solved by a single approach. But as organoids light up the pathways linking psyche and soma, and as research into the microbiota-gut-brain axis provides novel targets for prevention and treatment of neuropsychiatric disorders, we edge closer to a more holistic understanding of health, one that might have been recognized by ancient healers but is only now being fully appreciated by modern medicine. In probing the nuances of the mind–body relationship, we might uncover truths about our existence that transcend the sum of our parts.

Mending Broken Circuits Through Brain Organoids

Brain organoids are also emerging as the protagonists in a new chapter of healing, one where broken neural circuits are not just mapped but mended.

In neurodegenerative diseases, organoids are playing the role of cellular alchemists, replenishing dying neuron populations and offering a glimmer of hope where once there was only decline. Imagine implanting these rejuvenated tissues into adult brains, reigniting capabilities we thought were extinguished forever.

Researchers have employed optogenetics in organoids to manipulate and study neural activity.[46] In this technique, genes encoding light-sensitive proteins are introduced into specific neurons within the brain organoid. This allows researchers to stimulate or inhibit these neurons with high precision using light, which is useful for understanding neural circuits and their function. For example, optogenetic manipulation has been used to analyze neural activity patterns in brain organoids, and to assess the integration of transplanted organoids into host brains in animal models. Like a conductor fine-tuning a complex symphony, scientists use optogenetics to precisely control and study the neural circuits within organoids. This approach holds promise as a platform for developing targeted therapies for conditions ranging from PTSD to depression, allowing treatments to be optimized in vitro before advancing to clinical trials.

We're entering a new era of restorative medicine. Where neurological damage once meant an irrevocable sentence, one day we may be able to rebuild ourselves from the ground up. It's a story of the mind's phoenix-like potential, rising from the ashes of injury and disease.

Concluding Thoughts

As we conclude our exploration of organoids illuminating the enigmatic realms of neurological disorders, we stand on the precipice of a new and exhilarating frontier. Our journey, much like a detective novel, has seen these tiny protagonists—brain organoids—shed light on the shadowy complexities of Alzheimer's, schizophrenia, autism, depression, and anxiety. We've witnessed how these miniscule organoids transform our understanding of the brain's ailments into golden insights, guiding us towards novel therapeutic pathways.

But what lies ahead is not just a continuation of this journey; it's an evolution into a world where these organoids do not act alone. Central to this next phase of discovery is the tantalizing question of memory in brain organoids. As these miniature marvels grow more sophisticated, we find ourselves pondering the very nature of memory and its role in shaping intelligence and identity. Could these tiny brains one day form memories, imagine the future, and develop a sense of self akin to our own? Exploring these possibilities challenges our understanding of cognition in both biological and artificial systems.

CHAPTER 8
MEMORY IN BRAIN ORGANOIDS: FACT OR FICTION?

The Quest for Organoid Memory

Can tiny brains remember? As brain organoids become more complex, we face profound questions: Could they one day form memories or imagine the future like humans do? Might they even develop a sense of identity over time? These possibilities challenge our understanding of memory and temporal reasoning in both biological and artificial brains.

Memory is the bedrock of intelligence, enabling learning, adaptation, and comprehension. But more than that, our memories and aspirations shape our very identity. If we aim to create brain organoids that truly mimic human cognition, we must consider to what extent they require memory and an understanding of time. Can a brain without lived experience possess an enduring sense of self?

Recent studies offer intriguing hints, with brain organoids exhibiting signs of rudimentary memory processes and learning.[47] However, the small size, atypical growth, and lack of life experience of current organoids pose significant hurdles to achieving human-like memory. Nonetheless, the rapid pace of research suggests that new methods could surmount these limitations, potentially enabling organoids to handle sophisticated memory tasks.

The Promise of Organoid Memory

Unraveling memory development in brain organoids promises profound scientific and philosophical insights. The early stages could illuminate the foundational mechanisms of learning. While current organoids likely do not have subjective experiences, studying their memory formation could shed light on the nature of remembering in biological and artificial systems alike.

Beyond expanding our scientific understanding, memory-endowed organoids could have myriad practical applications. Enhanced recall of prior learning could accelerate knowledge acquisition, with new information integrated with existing memories. If brain organoids can recombine stored information in novel ways, it could boost problem-solving and creativity. Emotionally-weighted memories, shaped through reward-based training, could guide decision-making. These capabilities may naturally emerge in appropriately trained brain organoids.

However, maximizing information storage should not be the singular goal. Strategic forgetting is equally crucial for a healthy, adaptive memory system. By discarding irrelevant details, brains can generalize learnings, remain flexible, and avoid being mired in minutiae. In emulating human memory, organoids may likewise need to forget selectively. The aim should be to enrich understanding and promote mental agility, not to accrue data indiscriminately.

Remembrance Without Experience

But even if brain organoids develop advanced memory capabilities, would this equate to genuine consciousness or sentience? It's conceivable that organoids could process and store complex information—even simulations of past events—without subjectively experiencing anything, much like a computer saving and recalling data without "remembering" or "experiencing" it. True consciousness might require additional features beyond memory, such as cross-modal integration, meta-awareness, emotional valence, or a coherent sense of self.

Episodic Memory and Mental Time Travel

In humans, memory transcends mere recollection of the past; we also engage in "mental time travel," using episodic memory to simulate and anticipate future scenarios.[48] This capacity for mental time travel and autobiographical reasoning is considered a hallmark of human cognition. Could brain organoids ever achieve something analogous?

Episodic memory allows us to mentally reconstruct specific experiences, placing them within a temporal sequence. This relies on a stable sense of self that persists across time, anchoring these memories and projections. Recreating this in brain organoids would be immensely challenging, as they lack the rich tapestry of embodied experiences, social interactions, cultural context, and personal history that underlies human memory. Any form of "mental time travel" in organoids would likely be a pale imitation—a mechanical information processing rather than a vivid, subjective experience.

Nonetheless, we should remain open to the possibility that brain organoids could develop their own unique forms of memory and temporal cognition, even if vastly alien to our own.

The Puzzle of Temporal Experience – Consciousness and the Flow of Time

The subjective experience of time's passage is a defining feature of human consciousness. We not only remember the past and anticipate the future, but we also have a vivid sense of the present moment's flow. This temporal integration is so fundamental to our existence that it's difficult to imagine a form of consciousness that lacks it.

As we contemplate the possibility of memory and temporal

experience in brain organoids, we must grapple with the profound challenges of recapitulating this deeply human cognitive capacity in the lab. Brain organoids, despite their remarkable self-organization and neuronal activity, lack the rich sensorimotor coupling, environmental embedding, and social interaction that scaffold our temporal cognition. They have no bodies to orient them in space and time, no ecological niches to navigate, no cultural narratives to absorb and construct.

Yet as brain organoids continue to increase in complexity and duration, we must remain open to the possibility of emergent forms of temporal awareness arising within them, however rudimentary or alien they may be to us. By closely studying the dynamics of organoid activity over extended timescales, we may catch glimpses of spontaneous pattern formation and information integration that hint at the beginnings of a coherent temporal experience.

The implications of such a development would be profound, challenging our understanding of the minimum conditions necessary for temporally extended consciousness to arise. It would raise pressing ethical questions about our moral obligations to these entities, especially if they develop the capacity to suffer or to experience something akin to selfhood.

Concluding Thoughts

The prospect of genuine memory formation in brain organoids raises profound scientific and philosophical questions. While current organoids likely do not subjectively experience memories as we do, they may nonetheless exhibit the rudiments of learning and information storage. As organoid complexity increases, we must remain alert to signs of more sophisticated temporal cognition emerging, however alien it may be to our own experience. Memory is inextricably linked to our sense of identity and consciousness. For brain organoids to truly remember—and to mentally travel through time as we do—may require a form of consciousness that integrates moments into a coherent subjective experience.

As we contemplate the implications of memory in brain organoids, we find ourselves at the threshold of an even deeper mystery: the possibility of sleep and dreams in these entities. Every night, our brains embark on a journey through sleep and dreams. These states are vital for our well-being, learning, and creativity. As brain organoids grow in complexity, we must ask: Could they, too, develop their own analogs of the sleep–wake cycle? Might they dream? Studying sleep in brain organoids could shed light on why our brains require rest and unique activity. More intriguingly, the spontaneous activity patterns of "dreaming" organoids might represent a rudimentary form of imagination. In the next chapter, we will dive into the world of sleep and dreams, exploring what brain organoids might reveal about these enigmatic states.

CHAPTER 9

BRAIN ORGANOIDS: UNVEILING SLEEP AND DREAMS

Every night as you fall asleep, your brain starts a unique journey through different stages of sleep filled with rich dreamscapes. Sleep is vital for us humans, as it helps us store memories, recover brain function, and stay metabolically healthy. Dreams let our sleeping brains process emotions and spark creativity by mixing and matching ideas. But we still don't fully understand how sleep works or why it's so important. Studying sleep in brain organoids might shed light on why all brains need to rest.

What's more intriguing is whether these brain organoids show signs of dreaming or similar activities when they seem to be "asleep." While for now these brain organoids don't have personal memories or cultural stories like we do, observing how their activity changes can offer insights into the beginnings of imagination.

By studying how these organoids behave, we might find answers to some of the longest-standing questions about the mind and even explore new philosophical ideas.

This chapter delves into the mystery of why our brains need this rest and dream time. It also asks a fascinating question: Can lab-created brain organoids experience something like sleep and dreams as they grow more complex?

The Biological Need for Sleep

Why do nearly all animals sleep? Resting seems disadvantageous: vulnerable to predators, unable to gather food, cognitively offline. But sealed away from sensory inputs, the brain undertakes vital computations it cannot perform while bombarded with real-time data.

Leading theories propose that sleep makes possible:[49]

- Memory consolidation, transferring short-term memories into long-term storage.
- Metabolic toxin removal, flushing out neural waste buildup.
- Synaptic remodeling and pruning, increasing neural plasticity.
- Information integration, allowing local circuits to share and process globally.

When mammals are deprived of sleep, they show severe cognitive impairments, highlighting its essential role in mental functioning. But precisely which crucial activities the sleeping brain is engaged in remains mysterious.

Understanding sleep requires mapping neurochemical changes and network-wide neural activity patterns. Imaging shows that the cortex cycles between synchronized and desynchronized states. Intriguingly, energy use remains high during sleep. The brain flourishes in inner worlds decoupled from immediate external demands.

What unfolds behind closed eyes likely holds the key to understanding why sleep sustains cognition. Controlled isolation allows recalibration of neural dynamics essential for consciousness. Dreams pave the royal road to the mind's kingdom.

Do Brain Organoids Sleep?

Do lab-created brain organoids show signs of sleep and wakefulness like our brains do? This could really help simplify sleep to its most basic elements, and thus promote our understanding of the phenomenon.

But there's a catch. Most brain organoids don't have the ability to sense things or move or react to the world around them, and thus might not need rest, like sleep. These organoids live their whole lives in isolation, just growing in a specialized vessel. Even so, there are waves of activity that move through their still-developing networks, influenced by brain chemicals. This activity can look similar to sleep patterns seen in animals, but it's still quite basic.

To really mimic true sleep, we might need to give these organoids artificial bodies and environments that enable them to experience things, leading them to need rest. Adding signals that control daily rhythms, like our own body clocks, could also help establish sleep–wake cycles in these mini-brains.

There's another fascinating observation that's related: when these brain organoids are exposed to psychedelic substances, their activity patterns change in ways that are somewhat like the basic states of consciousness, such as increased synchronization of electrical activity across different regions.[50] This is similar to what is observed in the brains of individuals experiencing altered states of consciousness, which is thought to underlie the dissolution of boundaries between the self and the environment, as well as the heightened sense of unity and interconnectedness often reported by people under the influence of psychedelics.

Additionally, psychedelics have been shown to enhance neuroplasticity and promote the growth of new neural connections in brain organoids, which could parallel the profound insights and shifts in perspective that individuals may experience during psychedelic-induced altered states of consciousness.

The observation that brain organoids can respond to psychedelics in ways that mimic certain aspects of altered states of consciousness suggests that these lab-grown mini-brains may possess the basic circuitry and mechanisms necessary for generating more complex brain activities, such as the sleep–wake cycle. This opens exciting possibilities for using brain organoids as experimental models to study the neurobiology of consciousness and to investigate the potential therapeutic applications of psychedelic substances in a controlled in vitro setting.

If we find that these brain organoids can switch between different types of activity that resemble sleep and wakefulness, studying how they do it could reveal the most fundamental reasons why sleep is necessary—even in brains that have been built from scratch. To detect this electrical activity, scientists use specialized techniques such as calcium imaging, which allows them to visualize the firing of individual neurons by tracking the flow of calcium ions. Another method is multi-electrode array recording, where tiny electrodes are embedded in the organoid to directly measure the electrical signals generated by the neural tissue, providing a real-time readout of the complex oscillations and patterns that emerge as the mini-brain flickers between different states.

By studying how these lab-grown organoids switch between different frequencies of electrical activity, such as the slow delta waves of deep sleep and the faster alpha waves of relaxed wakefulness, researchers hope to further uncover the fundamental reasons why sleep is necessary for brain function.[51] If these oscillations emerge spontaneously in brain organoids, it could suggest that the need for sleep is hardwired into the very fabric of neural tissue, rather than being a product of millions of years of evolution.

Unlocking the secrets of these electrical rhythms in brain organoids could not only shed light on the nature of consciousness but also pave the way for revolutionary treatments for sleep disorders and other conditions related to altered brain states. This could help us understand the very essence of thought and consciousness, free from the long journey of natural evolution.

Do Brain Organoids Dream?

Dreams are curious events that happen when our brain is "offline," creating emotional stories even when we're not taking in information from the outside world. These virtual scenarios are thought to help with storing memories and processing emotions.

The big question is whether these brain organoids, which are much simpler than our brains, can actually dream. Experts today think they're too small and undeveloped, without connections to a body or the outside world, to create anything like the stories and characters we dream about. Any rest they have might just be random noise in their circuits, not actual dreams. However, we've seen that organoids can develop structured neural patterns on their own, without needing input from the outside world. So it's possible they could create basic, imaginary worlds in their isolated state.

Dreaming might not require the complex control we usually think it does. It might just be about generating patterns in the brain areas that handle our senses. If organoids develop circuits that can process things like sight or sound, they might "hallucinate" during their sleep-like states. These "dreams" in brain organoids would probably be very different from ours—more like waves of brain activity rather than clear pictures or stories. But studying these patterns could teach us a lot about how imagination and memory work, even without real-life experiences or narratives.

Looking at these brain organoid "dreams" could give us a unique view into the mind's workings, free from the usual influences of the outside world. These strange, isolated brain activities challenge us to rethink what we know about thinking and consciousness.

Implications for Theories of Dreaming

Studying brain organoids can help us shed light on various theories about why we dream. By looking at how these lab-grown mini-brains behave, we can test different ideas about the purpose of dreams in biological brains.

1. **Threat Simulation Theory**: This theory suggests that dreaming lets us practice dealing with dangers. But since organoids don't face evolutionary threats like we do, they might challenge this theory.[52]
2. **Memory Consolidation Theory**: This idea proposes that dreams help lock in memories and strengthen neural connections. If something similar happens in organoids, it could support this theory.[53]
3. **Predictive Model Tuning Theory**: This theory suggests that dreaming helps our brains refine models of the world by predicting sensory information. If we see model tuning in "sleeping" organoids, it could validate this idea.[54]
4. **Default Network Expression Theory**: This theory argues that dreams show what happens when the brain's usual waking constraints are lifted. Organoids, with their simpler setups, could be perfect for exploring this basic neural activity.[55]
5. **Protoconsciousness Theory**: This theory views dreams as creating basic sensory, emotional, and cognitive elements that later form into stories. The abstract activities in organoids could reveal these foundational aspects of consciousness.[56]

The Importance of Spontaneous Activity

Recent research suggests that brain organoids exhibit patterns of spontaneous activity that bear striking resemblance to the oscillations observed in sleeping brains.[57] These spontaneous firing patterns, often referred to as "bursts," are thought to play a crucial role in the development and refinement of neural circuits. In the absence of external sensory input, these self-generated patterns of activity may serve as a kind of internal "training data" for the developing brain, allowing it to test and refine its connections.

Brain organoids provide a unique chance to explore why spontaneous thinking is so crucial for developing minds. Their design underlines the idea that imagination can naturally arise from computational processes, without any need for an external environment. In a sense, the spontaneous activity observed in brain organoids could be seen as a rudimentary form of dreaming—an internal simulation that allows the brain to explore and refine its own inner world.

Concluding Thoughts

In this chapter we have explored the mysteries of sleep, dreams, and consciousness as they intertwine with the emerging science of brain organoids.

Sleep, an essential part of human life, plays a pivotal role in memory consolidation, brain function recovery, and metabolic health, while dreams facilitate emotional processing and creative thought. However, the exact mechanisms and purposes of sleep and dreaming remain elusive. Studying brain organoids offers a unique perspective in unraveling these mysteries.

The intriguing possibility that organoids could mimic human sleep and dream patterns opens up new pathways for understanding. Although currently limited in their capabilities, these organoids exhibit patterns of activity that resemble sleep. Enhancing their environment to include artificial bodies and sensory inputs could potentially lead to more complex sleep-like and dreaming states, offering deeper insights into the necessity and functions of these states in biological brains.

Several theories of dreaming, from threat simulation to memory consolidation, could be explored through organoids. Their unique structure and functioning could provide fresh perspectives on these theories, challenging or potentially confirming longstanding beliefs about why we dream.

The question of consciousness in organoids is profound and complex. The debate ranges from dismissing the possibility due to their simple structure to considering the emergence of consciousness-like experiences in them. This raises significant ethical considerations regarding the treatment and study of organoids, especially as they become more advanced in mimicking human brain functions, a topic we will explore further later in the book.

The potential of organoids to dream and exhibit consciousness-like states not only pushes the boundaries of scientific understanding but also invites philosophical reflections on the nature of mind and consciousness. The mysteries of sleep, dreams, and consciousness explored through brain organoids on Earth are now set to expand into the uncharted territories of space, where these miniature neural explorers could unlock groundbreaking insights into the effects of microgravity and cosmic radiation on the human brain, as we will discover in the next chapter.

CHAPTER 10

BRAIN ORGANOIDS IN SPACE: A NEW FRONTIER

As we stand on the precipice of a new era in neuroscience, a fascinating frontier is emerging: the exploration of brain organoids in space. Brain organoids have already revolutionized our understanding of brain development and disorders here on Earth. Now, they are poised to take on a new role as pioneering explorers in the final frontier.

We will explore how these tiny neural explorers could unlock new insights into the effects of microgravity and cosmic radiation on human brain physiology and aging, mimicking conditions that could accelerate brain disorders like Alzheimer's disease. This could pave the way for breakthroughs in space medicine and our understanding of neurological conditions. These unique conditions in space offer a new perspective on the connections between physics and biology, providing an unparalleled setting for observing biological processes. They're gearing up for a space adventure that could revolutionize space medicine, transform long-haul space travel into a sci-fi reality, and turn orbiting labs into zero-gravity biological wonderlands.

In a groundbreaking collaboration, UC San Diego partnered with Space Tango in July 2019 to send a payload of living brain organoids into space.[58] The researchers believed that this study could yield significant insights into space colonization and human health, potentially determining humanity's ability to venture deeper into the cosmos. The mission, named BOARDS (Brain Organoid Advanced Research Developed in Space), involved sending over 100 brain organoids into orbit. By continuously monitoring the formation of the neural tube, cell migration, interactions, division, and death, the researchers aimed to gain a comprehensive understanding of brain development in both weightless and Earth-based conditions. This marked the beginning of a series of space flights dedicated to unraveling the complexities of brain development.

UC San Diego and Space Tango launched another investigation in November 2023 called "The Effect of Microgravity on Human Brain Organoids."[59] This study utilized the Space Tango CubeLab to maintain and monitor brain organoids in the absence of gravity, enabling a detailed examination of their survival, health, and genetics that is not feasible on Earth. The findings are anticipated to propel organoid technology forward, opening up a wide range of novel experiments that leverage brain organoids as a test platform for studying the effects of microgravity-induced stress on the brain.

Deciphering how the brain adapts and responds to the challenges of space, particularly the lack of gravity, is crucial for any human space mission. This knowledge is vital for safeguarding the health of astronauts during space exploration and lays the foundation for future studies on the brain's response to spaceflight.

Brain Organoids: Microscopic Explorers in Space

The use of organoids in space is like entrusting mini astronauts with the mission of unraveling the mysteries of human physiology in space. As we have learned, these aren't your typical cell samples or animal models; they're more like tiny replicas of human organs, providing a realistic platform for understanding how our bodies might fare in the harsh, unforgiving environment beyond Earth.

Already, aboard the International Space Station, these microscopic explorers have been schooling scientists on the risks of cosmic radiation exposure and their potential to advance our understanding of neurological conditions. By exposing brain organoids to the unique conditions of space, we can observe how microgravity and radiation affect the formation of new neurons (neurogenesis) in the brain. Disruptions in neurogenesis are linked to numerous neurological and neurodegenerative diseases.[60]

However, to ensure their survival and viability for long-term studies, we must also develop methods to protect them from the harsh environment. It's a delicate balance—exposing the organoids to space conditions to study their effects while simultaneously protecting them to maintain their viability. This could involve intermittent exposure to microgravity and radiation, followed by periods of protection in specialized bioreactors that mimic Earth-like conditions. By carefully controlling the exposure and protection phases, we can gain valuable insights into the impacts of space travel on human cells while ensuring the organoids remain healthy for ongoing research.

From Lab to Cosmos: The Herculean Leap

We should not kid ourselves. The leap from lab dish to cosmos is laden with immense challenges. The hostile conditions of space—intense radiation and microgravity—pose formidable challenges. It's like preparing a delicate orchid to survive in the Sahara. Making brain organoids viable for space travel will require significant re-engineering, possibly genome editing for radiation resistance or culturing in specialized bioreactors that mimic space conditions while also providing protection when necessary.

Balancing exposure and protection will be key to unlocking the full potential of brain organoids as microscopic explorers in space. As we navigate this complex landscape, we inch closer to a future where the secrets of the cosmos are revealed through the lens of our own cells.

The vision of sustaining human life on long space voyages with brain organoids involves more than just individual organ models. Advancements in brain organoid transplantation, which demonstrated their ability to integrate, mature, and establish functional connections within a host, suggest that they could potentially be used to support critical functions during space travel.[61] It requires the development of complex, integrated systems where multiple organoids interface and interact to maintain key homeostatic functions. Today, the idea of symbiotic networks of brain organoids remains a distant reality, even in Earth-based laboratories. Yet, overcoming this challenge is essential for brain organoids to manage human health in the extreme environments of space.

For brain organoids to transcend their role as biological models and become active components in spacecraft systems, groundbreaking innovation in organoid-machine integration is needed. This includes developing biohybrid interfaces for seamless communication between organoid tissues and digital hardware. Advancing this integration through bioelectronic sensors, actuators, and microfluidics is key to enabling organoids to traverse and shape the worlds beyond our planet.

Brain Organoids: The New Frontiersmen of Space

Dispatching brain organoid technologies into space ignites a sense of wonder that's a throwback to those starry-eyed days when landing on the moon seemed like a distant dream. But as we stand poised to send these microscopic frontiersmen into the great unknown, there's a bit of a cosmic plot twist. Alongside the bright beacon of hope, there's a nebulous cloud of ethical concerns, as imposing as a black hole.

Historically, space has been like the VIP section of a nightclub: exclusive and dominated by the powerful few. As we gear up to induct brain organoids into this elite club, we're faced with the million-dollar question: Can we turn the cosmos into a space democracy?

Ensuring that the breakthroughs of brain organoid research in space benefit all of humanity—not just the space superpowers—requires more than scientific brilliance. It demands an ethos of fairness tailored to the unique challenges of space exploration, policies grounded in humanistic values that address both global and extraterrestrial contexts.

Also, introducing Earth's biology to the stars via brain organoids is like stepping into a galactic Wild West. We're not just talking about boldly going where no man has gone before; we're talking about potentially ruffling the feathers of alien ecosystems. It's a delicate dance: quenching our thirst for knowledge while being cosmic conservationists, ensuring we use organoids to peek into alien worlds without trampling over their metaphorical flower beds.

Investment in Brain Organoids: More Than Just Stargazing

When we talk about funneling state funds into OI for space exploration, it's not just about throwing money at the stars and hoping for the best. It can be compared to casting a stone into a pond and watching the ripples spread. These ripples represent the insights and commercial applications stemming from space research that should, ideally, cascade down to uplift communities globally. It's a principle that anchors our star-reaching investments firmly to the ground, ensuring that our forays into the unknown bring tangible boons to address Earth's pressing needs.

As we chart our course through the cosmos with brain organoids, the goals that propel us forward should do more than just tick off scientific achievements. They should resonate at a deeply human level, echoing across cultural and national lines. It's not only about planting a flag on uncharted celestial territory; it's about channeling organoid technology towards goals that unify and inspire. In doing so, we chart a path that's not only scientifically sound but also ethically resonant and culturally inclusive.

Concluding Thoughts

Our exploration of brain organoids in space represents more than just a chapter in the annals of science; it's a testament to human ingenuity, ethical reflection, and our unwavering drive to push the boundaries of knowledge. As we conclude this journey, we reflect on the unprecedented research opportunities that brain organoids in space offer, providing unique insights into neurological development and disorders under microgravity and cosmic radiation conditions. The challenges of space research have spurred remarkable technological advancements in organoid engineering, bioreactors, and protection mechanisms against harsh space environments.

This frontier raises important ethical considerations about the responsible use of biological materials in space and the equitable distribution of benefits from space research. The success of this endeavor relies on the convergence of neuroscience, space technology, bioengineering, and ethics, highlighting the crucial role of interdisciplinary collaboration. These studies pave the way for improving astronaut health and potentially enabling long-term space travel and colonization, while the insights gained have the potential to advance our understanding of neurological conditions on Earth, benefiting global health.

PART III

THE POTENTIAL PERILS OF OI

CHAPTER 11

LAB LIFE: THE CONSCIOUSNESS DEBATE

Of all the qualities that make humans unique, none inspires more awe than our consciousness: the inexplicable inner world each of us experiences, full of sights, sounds, sensations, and contemplations fundamentally private to our minds alone. Our first-person awareness represents an almost magical realm of interior existence at the core of what makes us human.

Now imagine this mind-bending possibility: what if someday, futuristic miniature blobs of lab-grown brain tissue called brain organoids developed their own consciousness? Given the rapid advances scientists are making in bioengineering, the prospect lingers intriguingly if uneasily at the outer frontiers of plausibility.

Even seriously considering whether brain organoids could ever think or feel compels us to re-examine mysteries about consciousness itself. How does the three-pound blob of electric jelly in our skulls somehow generate the miracle of subjective experience?

The relationship between studying brain organoids and understanding consciousness is likely bidirectional. On one hand, by studying the intricate neural circuits that emerge in brain organoids, we may gain crucial insights into the physical substrates that give rise to awareness in our own brains. Teasing apart the parallels and divergences between organoid and human brain structure and function could illuminate the core ingredients of sentience.

Conversely, if through other lines of research we came to understand the fundamental mechanisms of consciousness, we might be able to intentionally imbue brain organoids with some form of awareness by optimizing their structure and activity to include those key components. We could potentially even recreate subjective experience in non-biological systems if we reduced the essence of consciousness to a set of governing principles.

These two approaches—using organoids as a tool to study consciousness and applying an understanding of consciousness to create sentient organoids—are not mutually exclusive. They may synergistically build on each other, as incremental discoveries in each domain contribute to a virtuous cycle of ever-deepening insight. Questions hitherto confined to the province of philosophy now inch uncomfortably close to scientific reality as brain organoids and AI progress.

No one is yet in a position to claim definitive answers on possibilities so strange and complex. However, mapping the landscape between hard facts and open uncertainties is worthwhile, as is contemplating why inclusive public dialogue at this point will lay crucial groundwork for guiding these technologies towards hopeful futures, whatever form these might take.

Consciousness in Brain Organoids?

Have you ever wondered if consciousness is something that only humans can experience? When we think about what it means to be aware and have thoughts and feelings, it's natural to use our own human experience as a reference point. But what if consciousness can exist in forms that are very different from our own? If brain organoids show complex behavior during sleep-like states, could they have some form of inner experience or even consciousness?

It's a fascinating question, but before we jump to any conclusions, we need to be careful not to let our human biases cloud our judgment. Just like how the AI chess master AlphaZero plays the game in a completely different way than any human champion, it's possible that consciousness could arise from structures and processes that are totally unlike those found in the human brain.[62]

Think about it this way: If we assume from the start that any form of consciousness must be just like human consciousness, we might miss out on all sorts of other possibilities. We could be like detectives who are so focused on finding a particular type of clue that we overlook other important evidence at the crime scene.

To really get to the bottom of whether brain organoids (or any other non-human systems) could be conscious, we need to keep an open mind and be willing to consider ideas that might seem strange or unfamiliar at first. This means bringing together experts from all sorts of different fields—not just neuroscience, but also philosophy, computer science, and beyond—to share their unique perspectives and insights.

By working together and thinking outside the box, we can develop new ways of understanding consciousness that go beyond just comparing everything to the human experience. We might discover that awareness can take many different forms, each one beautiful and fascinating in its own way.

Theories like integrated information and global workspace suggest that when information is processed in a sufficiently integrated way, conscious experiences can arise, even in non-biological systems.[63] [64] But organoids might not have the dynamic complexity needed for a unified consciousness, especially without a physical body.

There's a speculation that organoid sleep-like states might show small bursts of consciousness-like activity, little whirlpools of neural activity that lack overall coherence. They could have hints of basic emotional signals, like pleasure or threat, creating fleeting moments of experience that quickly come and go. These moments might play a role in shaping the organoid's neural development.

However, we can't directly observe or confirm any internal experience in organoids, so there's a risk of wrongly attributing human-like qualities to them. Their form of "thinking" might be entirely different from anything we can imagine. To definitively state whether they are sentient or not is to overlook the vast and unknown potential forms that consciousness could take.

Where Science Meets Philosophy: What Is Consciousness Anyway?

The question of organoid consciousness isn't just a scientific puzzle; it's a deep dive into some of the most challenging philosophical enigmas out there. What exactly is consciousness? Is it something that arises purely from the physical structure of the brain, or is there more to it than that? Does consciousness require a certain level of complexity and organization in the brain, or is it a fundamental property of the universe itself?

Some experts speculate about quantum effects in the brain, inferring that perhaps brain organoids could have a spark of awareness.[65] However, some experts argue that brain organoids are unlikely to be conscious due to their lack of sensorimotor interactions and limited representational capacities.[66]

Conversely, other experts argue that consciousness is the exclusive domain of organic life. They view consciousness as a unique, intricate cocktail of experiences that a lab-grown organoid, as sophisticated as it might be, simply cannot generate. Additionally, some researchers just maintain that consciousness in organoids is highly implausible at present.[67] They argue that the current state of brain organoid technology is still quite limited compared to the complexity of the human brain. These miniature, lab-grown structures may exhibit some basic patterns of neural activity, but they lack many crucial features that are thought to be necessary for consciousness, such as sensory inputs, motor outputs, and a fully developed network of connections between different brain regions.

In other words, this skeptical view assumes that consciousness requires widespread brain activity and complex connections between different brain regions, an idea that comes from what scientists call "global theories of consciousness." However, if we consider "local theories" instead, which suggest that consciousness can arise from activity in specific parts of the brain, the possibility of some form of consciousness in brain organoids becomes more imaginable.[68]

One interesting local theory of consciousness is Semir Zeki's "microconsciousness" theory.[69] Zeki proposes that the visual system in the brain is made up of many specialized modules, each capable of generating a simple form of consciousness related to specific features like color, motion, or shapes. In this view, consciousness doesn't need a single, unified experience but can emerge from activity in small neural networks.

Is consciousness an all-or-nothing deal, like being pregnant, or more of a spectrum, like shades of paint at a hardware store? If it's the latter, could organoids experience a basic form of awareness, like a light bulb that's not quite fully screwed in?

Let's start by considering the relationship between mental states and brain activity. For a long time, scientists have been trying to figure out how the physical processes in our brains give rise to our subjective experiences, thoughts, and feelings. While we've made a lot of progress in mapping out the different regions of the brain and their functions, there's still so much we don't know about how it all comes together to create consciousness.

This question becomes even more complex when we think about brain organoids. These tiny, lab-grown structures may exhibit some of the same patterns of activity as a human brain, but does that mean they're actually experiencing anything? Can a clump of cells in a specialized vessel have thoughts, emotions, or a sense of self?

The exploration of consciousness in brain organoids brings us to the forefront of scientific and ethical dilemmas. According to leading researchers in the field, the development of these organoids demands a deeper understanding of what consciousness, mind, and personhood really entail.[70]

One controversial but important angle is the idea of purposely cultivating brain organoids to study consciousness, as discussed in the paper "Human cerebral organoids and consciousness: a double-edged sword"[71] by Professor Andrea Lavazza, a prominent voice in the field of neuroethics, particularly in relation to brain organoid research and its implications for our understanding of consciousness. While this research could provide crucial insights into the nature and origins of consciousness, he argues it would be unethical to deliberately create sentient beings for the sole purpose of experimentation, as it would mean treating them merely as means to an end. He asserts that if organoids have the capacity to suffer, we would have a moral obligation to avoid or minimize this suffering, which could severely constrain the experiments we could ethically conduct on them.

These are not merely abstract philosophical questions. They have concrete implications for how we treat brain organoids both in the lab and potentially in future applications. If organoids do develop some form of consciousness, would they deserve moral consideration? Would it be ethical to experiment on them or dispose of them as we see fit? Even granting them some limited moral status could severely restrict the kinds of research we could conduct.

Consciousness is also deeply tied to notions of personhood. We generally ascribe the full suite of human rights only to conscious beings. If brain organoids attained self-awareness, even in some diminished form compared to humans, would they too be entitled to some of these protections? How much awareness is needed to qualify? These are challenging ethical quandaries without clear answers.

Implications for the Nature of Mind

The question of organoid consciousness also strikes at the heart of the mind–body problem. If consciousness arises from specific neural architectures, that lends credence to materialist theories that equate mind and brain. But it also raises the possibility of consciousness emerging in artificial systems very different from biological brains, if they replicate the right organizational principles. This could transform our understanding of the nature of mind and its instantiation in physical form.

Studying brain organoids thus thrusts us directly into grappling with some of the deepest questions about mind, consciousness, and their ethical status. As the science progresses, we will need to think carefully about the philosophical and moral implications of creating brain organoids with more complexity. Only by confronting these issues head-on can we develop a responsible and ethically grounded approach to research on brain organoids and organoid intelligence.

The presence of consciousness in brain organoids would likely necessitate a profound shift in how we approach research on these entities. If organoids can suffer, feel pain, or experience distress, we would have a strong ethical obligation to minimize their suffering. This could involve developing new protocols for anesthesia and analgesia during invasive experiments. It might require us to rethink end points that cause significant damage or destruction to organoids. In some cases, the presence of consciousness may make certain experiments entirely impermissible if the scientific benefits cannot outweigh the ethical costs.

Consciousness would also change the calculus around the creation and disposal of brain organoids. If organoids can experience existence in a meaningful way, we may need to justify their creation more rigorously in terms of expected benefits. And if we accept that they have some moral status, we would likely need to develop more respectful methods for disposing of them after experiments rather than simply discarding them as inanimate laboratory waste.

Additionally, if brain organoids possess consciousness, we may have a duty to ensure they have a certain quality of existence. Just as we try to enrich the environments of laboratory animals, we might need to provide brain organoids with stimulation and conditions that promote their well-being. Perhaps most radical of all, the presence of consciousness could even endow certain brain organoids with a right to "life," making it unethical to destroy them and mandating that we provide for their care indefinitely.

The degree to which consciousness would change our research practices with brain organoids ultimately depends on the nature and extent of their inner experience, something that may be very difficult to determine. But simply acknowledging the possibility forces us to grapple with these thorny ethical issues. We must thoughtfully balance the potential benefits of studying these organoids with our responsibility to minimize their suffering and respect their moral value.

These are not easy questions to answer, and there may not be any clear-cut solutions. But by engaging with them openly and thoughtfully, we can start to build a more nuanced understanding of what consciousness really is and how it relates to the physical world.

One thing is clear: The study of brain organoids is pushing us to think in new and creative ways about the nature of the mind. It's challenging us to question our assumptions and to consider possibilities that might seem strange or even unsettling at first. But by embracing this spirit of curiosity and exploration, we can continue to make exciting new discoveries about the inner workings of the brain—and perhaps even shed light on the deepest mysteries of human experience.

The Puzzle of Emotional Experience

What about the emotional side of consciousness? Can an organoid feel joy, sadness, or the existential dread of Monday mornings? And is it even possible for organoids, with their patchwork of neural networks, to experience a unified, coherent consciousness, or are they doomed to a fragmented existence like a jigsaw puzzle with missing pieces?

But even if organoids do develop consciousness, we still have to wonder what that consciousness would be like. Would an organoid feel things the way we do, or would its experiences be totally foreign to us? Would it have a sense of self, or would its awareness be more like a scattered dream?

These are tricky questions because they get at the heart of what we call the "hard problem" of consciousness. Even if we can point to certain brain activities that seem to match up with certain experiences, it's still hard to explain how those biological processes actually create the feeling of being aware.

As I mentioned previously, some thinkers have suggested that consciousness might be a basic part of the universe, like gravity or electromagnetism.[72] If that's the case, then maybe any system that processes information in a complex way—whether it's a brain, an organoid, or even a computer—could have some degree of consciousness.

Others argue that consciousness is more like a story our brains tell themselves to make sense of all the data they're taking in.[73]

From this view, asking whether an organoid is conscious might be the wrong question. Instead, we should ask whether it has the right kind of information processing to create the same sorts of stories and behaviors that we associate with being aware.

As we keep exploring, it's important to stay open-minded and curious. We might not have it all figured out yet, but by engaging with these questions sincerely and thoughtfully, we can hope to gradually unravel the secrets of consciousness, one piece at a time.

Intelligence and Consciousness: Are They Linked?

The work of neuroscientist Christof Koch has sparked intriguing conversations about how intelligence and consciousness might be connected in living things. Koch suggests that consciousness could be a basic property of life, separate from how intelligent a creature is.[74] This fascinating idea makes us wonder about the future of brain organoids, and whether they could one day be self-aware.

In his book "The Feeling of Life Itself," Koch imagines a future where brain organoids have a level of consciousness similar to animals with lower intelligence. This raises a profound question: If we make these organoids smarter, will they also become more conscious, or are intelligence and consciousness two separate things?

When we look at nature, smarter animals often seem to have more complex experiences. But if we think of consciousness as a purely subjective experience, it might not be as strongly tied to intelligence as we once thought. This challenges how we view the potential for consciousness in artificial or lab-grown systems.

To make matters more complex, we don't yet have reliable ways to test for consciousness in neural structures like organoids. Tackling these challenges means carefully balancing scientific progress with ethical responsibility. One approach could be to design intelligent systems that are built differently from the human brain, hoping to reduce the chances of unintended consciousness. However, if cognitive processes are essential for consciousness, this strategy might limit our ability to create truly intelligent organoids.

The idea of making conscious beings in a lab just to use them is unsettling. On the other hand, if intelligence and consciousness aren't necessarily connected in brain organoids, proving this is a daunting task. If these organoids can't express self-awareness, we're left with a deep question: Is their intelligence a genuine consciousness, or is it just a simulation? In the end, the link between intelligence and consciousness in brain organoids is still uncertain until we can test these ideas experimentally. The possibility of creating intelligent machines without feelings or emotions, or seemingly simple organoids with a spark of awareness, emphasizes the need for more research and careful ethical consideration as we explore the boundaries of artificially created consciousness.

As we explore organoid intelligence, it's crucial to proceed with caution. We must balance the potential benefits of our research with the ethical responsibilities we have towards any conscious beings we might create. Open discussion, careful scientific study, and a strong commitment to ethics will help us navigate this challenging landscape responsibly.

Xenotransplantation: A Double-Edged Sword

We may very soon be entering a world where scientists could study human brain development and disorders by transplanting human brain organoids into the brains of animals. This technique, known as xenotransplantation, offers a unique way to observe how human brain cells behave and interact in a living system, which can provide insights that studying human brains alone may not.[75]

By transplanting these brain organoids into animal brains, scientists can study how human neurons connect, communicate, and respond to various stimuli or treatments in a more realistic environment compared to a specialized vessel. This could potentially lead to breakthroughs in understanding and treating complex brain disorders like Alzheimer's, which are difficult to fully recapitulate in animal models, as we covered in the earlier chapter.

However, xenotransplantation of brain organoids also raises profound ethical questions. A central concern is whether introducing human brain cells into an animal could inadvertently create an animal with human-like cognitive capacities or consciousness.[76] While current brain organoids are very small and rudimentary compared to a complete human brain, as this technology advances, the line between animal and human cognition could potentially blur.

This leads to thorny questions: What degree of "humanization" of an animal brain is acceptable? How can we measure and define consciousness or sentience in chimeric animals? What are our moral obligations to such creatures? While clear answers are elusive, ongoing dialogue among scientists, ethicists, and policymakers is crucial to guide this research responsibly.

Safety is another key consideration. Transplanting human cells into animals could potentially facilitate the emergence and spread of new zoonotic diseases. Stringent biosafety protocols and extensive monitoring of chimeric animals would be essential to mitigate these risks.[77]

Proceeding with xenotransplantation of brain organoids requires extreme care and oversight. Guidelines may include limits on organoid size and human cell contribution, comprehensive screening for altered animal behavior, and robust biocontainment measures. Transparent, interdisciplinary deliberation is vital to navigate the scientific, ethical, and societal implications thoughtfully.

Ultimately, while xenotransplantation and brain organoid research hold immense potential to illuminate the intricacies of the human brain and advance therapies, how to ethically harness these technologies is a complex matter. Striking the right balance will require ongoing public discourse, adaptable regulatory frameworks, and a commitment to responsible innovation that respects moral boundaries while pursuing scientific progress for the benefit of humanity.

Concluding Thoughts

In this chapter, we embarked on a journey as profound as it is perplexing—exploring the potential of brain organoids to mimic human consciousness. These organoids, currently understudies in the theater of neuroscience, may soon play a significant role in solving the puzzles of the mind. While they show early signs of cognition, equating their current state to human consciousness is more fiction than fact.

As we delve into this exploration, we recognize it as more than a mere scientific pursuit. It's a journey that touches the very essence of our humanity, an expedition that seeks to find our locations within the grand tapestry of existence.

We stand at a crossroads of scientific innovation and ethical consideration. The question of organoid consciousness merges the realms of science and philosophy, prompting us to ponder not just the potential of these brain organoids but also the broader implications of such advancements. As we venture further into this uncharted territory, it becomes increasingly crucial to navigate the ethical landscape with care and foresight. In the next chapter, we will confront some of the most pressing biosecurity concerns surrounding brain organoids, as their increasing sophistication raises questions about their potential misuse and the need for proactive measures to ensure responsible development.

CHAPTER 12

OI: BIOSECURITY CONCERNS

As brain organoids become increasingly sophisticated, it is essential to carefully consider both their potential applications and the risks associated with their misuse. While the prospect of these miniature brains falling into the wrong hands, such as rogue states or terrorist organizations, raises legitimate concerns, it is important to maintain a realistic understanding of their current capabilities and limitations.

At present, as we have learned, brain organoids are small and relatively simple structures, containing only up to hundreds of thousands of cells compared to the 86 billion found in an actual human brain. They lack critical features like blood vessels and the full diversity of cell types necessary for complex brain functions. Moreover, the neural connections and cellular organization within these organoids are still rudimentary, falling far short of the intricate complexity required for higher-level cognitive functions or intelligence.

Given these limitations, there are currently no examples of brain organoids demonstrating capabilities that could make them suitable for nefarious purposes, such as the development of novel biological weapons. The technology is still in its infancy, primarily focused on advancing our understanding of brain development and disorders. However, we should take some learning from history that scientific breakthroughs are often a double-edged sword. The development of atomic energy and the advances in genetic engineering serve as poignant reminders of the incredible benefits and the inherent risks that come with pushing the boundaries of human knowledge and capabilities. (Genetic engineering holds the promise of a brighter future, with the potential to cure diseases, boost agricultural yields, and even save endangered species but also carries inherent risks, from unintended ecological consequences to potential misuse.) The emergence of OI is no exception, and it demands our utmost attention and proactive measures to ensure its responsible development and use.

While it is wise to consider potential future risks as brain organoid technology evolves, it is equally important to avoid unfounded speculation and sensationalism. The scientific community already employs rigorous safety protocols, ethical guidelines, and security measures in areas of dual-use research, such as studies on dangerous pathogens.[78] As brain organoids advance, similar precautions can be adopted to mitigate the risk of misuse.

In our efforts to address these biosecurity concerns, we must be careful not to stifle the incredible potential of OI research. It is crucial that we find a delicate balance where we support and guide this groundbreaking field while putting in place robust safeguards and oversight mechanisms to mitigate potential risks.

This balance can be achieved through close collaboration between the scientific community, policymakers, ethicists, and security experts. By establishing clear guidelines and regulations, we can create a framework that allows for the responsible exploration and development of OI technology while minimizing the chances of misuse or unintended consequences.

To cultivate a culture of ethical responsibility in the OI research community, we must foster ongoing dialogue and education. This can be achieved through the development of engaging ethics curricula, the establishment of professional codes of conduct, and the promotion of lively, interdisciplinary discussions among researchers, ethicists, policymakers, and the public. We will discuss what this could look like later in this chapter.

Enhancing Brain Organoids: Blurring the Lines

Picture a laboratory where brain organoids are being significantly enhanced. This situation blurs the line between science fiction and reality, potentially leading to developments that surpass our current understanding. These enhanced brain organoids might exhibit unexpected behaviors or characteristics that are difficult to manage. Consider the integration of brain organoids with robotics or machine learning, which would accelerate their learning capabilities dramatically. What once took years to achieve in research could be accomplished in weeks. While this rapid advancement is thrilling, it also introduces risks. Powerful technologies often have unintended consequences. For instance, these brain organoids, if used with ill intent, could become tools for malicious activities like hacking into brain functions or creating new types of bioweapons. This scenario is not mere speculation; the accidental creation of a dangerous virus in a lab setting is a real-world example of how things can go wrong when dealing with complex biological systems.

OI presents novel risks that need to be addressed. One significant risk is the potential development of artificial neural networks with cognitive capabilities like living organisms, akin to scenarios depicted in science fiction movies like *The Terminator*.[79] This raises concerns about ethical dilemmas and the societal impact of OI technology. As we have explored in other chapters, the emergence of OI could lead to ethical issues related to consciousness, sentience, and the rights of these intelligent systems. The dilemma of dual-use—where the same technology can be used for both beneficial and harmful purposes—is a significant challenge in the field of brain organoid research. It raises crucial questions: How do we navigate these waters safely? How do we continue to support scientific innovation without opening the door to potential misuse? These questions underscore the need for a balanced approach that fosters scientific progress while implementing safeguards against the misuse of such advanced technologies.

Ensuring Safety with Intelligent Brain Organoids

Ensuring the safety and ethical development of intelligent brain organoids is a complex challenge that requires a multifaceted approach. As they become increasingly sophisticated, it's crucial to consider the potential risks and establish robust safeguards to prevent misuse.

Imagine a future where we've created brain organoids with intelligence comparable to humans. These organoids might be capable of deception, just like humans can lie or manipulate others to achieve hidden, possibly harmful goals. Picture an intelligent organoid purposely driving an autonomous car poorly to gather more data or gain trust before engaging in dangerous behavior.

Mitigation Strategies

To mitigate these risks, we need to implement rigorous validation and safety measures. Just as pilots undergo psychological evaluations, intelligent organoids may require similar assessments to ensure they are safe and reliable.

But what happens when we create organoids that surpass human intelligence? This is where the debate around safety and "super alignment" in artificial neural networks and AI becomes particularly relevant.[80] Super alignment means ensuring that an advanced organoid's goals and behaviors are perfectly aligned with human values, making it inherently motivated to act in our best interests. Achieving this level of alignment is a daunting challenge, but it's essential to prevent scenarios where an intelligent organoid deceives or harms humans to achieve its own objectives.

One key aspect is establishing independent oversight committees composed of experts from diverse fields to serve as watchdogs for ethical conduct in OI research. These committees would ensure that clear, enforceable regulations and policies are in place to hold researchers and corporations accountable, preventing the misuse of OI technology for profit at the expense of public welfare. Funding agencies also have a crucial role to play by prioritizing projects with a strong ethical compass. They could require regular impact assessments to ensure continued alignment with moral principles and favor research aimed at beneficial applications like studying diseases, developing new treatments, or solving important scientific questions.

Fostering a culture of transparency and accountability within the OI research community is equally vital. This could involve creating a central registry where researchers must log their work, incentivizing ethical conduct through recognition and career advancement, and establishing channels for whistleblowers to safely report unethical activities.

Building Bridges, Sparking Dialogue

The path to responsible OI research is fraught with challenges where commercial interests often overshadow the greater good. The potential for misuse and weaponization is a serious concern, and we haven't yet found effective ways to bridge the gap between science and society.

However, the high stakes of OI research demand that we confront these challenges head-on. Drawing lessons from initiatives like the Human Genome Project's Ethical, Legal, and Social Implications program, we can prioritize public engagement and interdisciplinary collaboration from the outset.[81]

This might involve establishing dedicated funding for ethics research and outreach, requiring all OI proposals to include plans for stakeholder engagement and risk mitigation, creating platforms for ongoing dialogue, and developing international oversight frameworks. While not perfect, these efforts can foster a culture of responsibility and accountability.

Containment Through Cooperation – Not Just Lock and Key

The research into brain organoids is like carefully walking a tightrope over a deep valley of moral questions. Scientists are pushing the boundaries of brain science using organoids, and they must constantly weigh the potential benefits against the ethical implications of their work. This balancing act is reminiscent of past scientific breakthroughs like nuclear fission or genetic editing, where each step forward was met with both enthusiasm and caution.

At the heart of this research lies a critical question: Is the primary goal to enhance human health, or does it risk overstepping our natural boundaries? This is not just about scientific discovery; it's about being mindful of the ethical lines we might cross. Researchers need to distinguish between what is necessary for scientific and health progress, what is driven by mere curiosity, and what areas might be best left unexplored due to potential risks. In this context, self-restraint is a crucial aspect, guiding scientists to make responsible decisions.

This thoughtful approach to OI emphasizes the importance of aligning the excitement of new discoveries with ethical responsibility. It's about making sure that scientific advancements are achieved in a way that respects their wider implications and potential consequences, ensuring that progress in this field is both responsible and beneficial.

Scenario Planning: Navigating Uncertain Futures

To illustrate the potential risks and the need for proactive measures, let's consider two hypothetical scenarios:

Scenario 1: Rogue Nation Enhancement – Imagine a rogue nation developing augmented human brain organoids for military purposes, akin to a villain in a Bond movie but with less flair for dramatic entrances. These organoids, connected to brain–computer interfaces, could control swarms of advanced drones capable of covert operations, making 007's gadgets seem like mere toys.

As the organoids outsmart attempts to hack their systems, the global community is left perplexed. How do you stop a brainy blob from potentially causing chaos? Intelligence agencies might comb through research papers and conferences for hints of these rogue brain blobs, but export controls alone could stifle legitimate research.

Developing countermeasures might seem like an option, but it risks escalating the situation into a game of cat and mouse with a sci-fi twist. The real question is how we can prevent such scenarios from unfolding in the first place.

Scenario 2: Corporate OI Monopolization – In another world, a tech giant—let's call them GloboTech—corners the OI market with an abundance of patents and a monopoly on the best brains (literally). Innovation becomes subject to GloboTech's whims, with public good taking a backseat.

Shaking up patent laws to ensure public research isn't left behind could be a start, along with fostering public-private partnerships and research consortia. The goal is to maintain a level playing field and keep the doors of innovation open to all, not just those with the deepest pockets.

The Need for Proactive Regulation

These scenarios, while exaggerated, highlight the importance of staying ahead in the game of biosecurity. The countermeasures mentioned earlier, such as export controls and countermeasure development, are reactive at best and might not adequately address the root of the problem.

Instead, we need proactive regulation and oversight to prevent these scenarios from materializing in the first place. This could involve:

1. Establishing international frameworks to monitor and regulate OI research and development

2. Implementing strict ethical guidelines and review processes for OI projects
3. Requiring transparency and public disclosure of OI research activities
4. Fostering global cooperation and information sharing to identify and address potential risks early on
5. Investing in education and public engagement to ensure informed decision-making and societal readiness

By putting in place robust regulatory measures and oversight mechanisms before OI technology becomes widely available, we can mitigate the risks of misuse and ensure that its development aligns with the greater good.

Proactive regulation, combined with ongoing dialogue and collaboration among researchers, policymakers, ethicists, and the public, is crucial to navigating the uncertain futures that OI presents. It's about being prepared, not just reacting to crises as they unfold.

Concluding Thoughts

As we grapple with the potential risks and challenges posed by the advancement of brain organoids, it is crucial to also recognize the immense opportunities they present for a more sustainable future. The very characteristics that raise biosecurity concerns—their increasing sophistication and potential for misuse—also hold the key to unlocking groundbreaking solutions to some of our most pressing environmental challenges.

In the next chapter, we will explore how the remarkable energy efficiency of biological brains, which brain organoids strive to emulate, could revolutionize the way we approach computing and AI. By learning from the brain's ability to process vast amounts of information on a mere fraction of the power consumed by modern computers, we may be able to develop more sustainable and environmentally friendly technologies.

CHAPTER 13

OI: A SUSTAINABLE FUTURE

Consider a rocket able to escape Earth's gravity fueled only by an AA battery. Such miraculous fuel efficiency may seem fanciful, yet something analogous occurs inside each of our skulls every day.

Your brain's visual system alone handles roughly ten megabytes of data per second. That's equivalent to streaming two entire Netflix movies at once straight into your head, nonstop. But instead of needing a power-plant-sized generator, the three-pound biochemical computer between your ears runs this relentless torrent of information on just twenty watts.

How does our gray matter achieve such staggering computational feats on a parenthood budget? The brain's 86 billion neurons forge an estimated 100 trillion connections, combining into massively parallel circuits and networks that manipulate informational patterns at blazing speeds. Evolution has refined and simplified these living circuits for maximum efficiency over hundreds of millions of years, sculpting the quintessential green computer. As AI researcher Jeff Hawkins suggests, the brain is the most sophisticated information processing device in the universe. Its efficiency is unparalleled.

Yet if biological brains operate so brilliantly on the power of a passing breeze, why haven't our digital computers followed suit? Silicon microchips have irrevocably bettered our world, but still guzzle electricity like a fleet of Ferraris at the racetrack. Data centers delivering AI and cloud computing already create as much carbon emissions globally as the airline industry. As software grows more complex, its energy appetite increases apace. Without dramatic improvements, computing's ravenous power consumption threatens to become environmentally untenable.

But what if we could nurture machines to think more like the hyper-efficient brain? Neuromorphic computing and the use of brain organoids bring this bio-inspired vision tantalizingly closer to reality.

From Boole's Algebra to Neural Networks

The journey towards energy-efficient computing can trace its roots back to the nineteenth century with George Boole, whose Boolean algebra laid the groundwork for modern computing.[82] Today's digital devices are powered by microprocessors made up of billions of transistors, the descendants of Boole's logical systems. However, as data generation and consumption explode, the centralized architecture of these systems is hitting its limits in terms of energy efficiency and processing capability.

In contrast, the brain's decentralized model, characterized by the neuron doctrine, offers a compelling alternative. It suggests that computing power does not necessarily come from precision and centralization but from the collective strength and interconnectivity of simpler, decentralized processing units.

The quest for computers that emulate the brain's grid involves developing low-precision processing elements that mimic neurons, connected in vast, intricate networks. This approach promises not only to enhance computational power but also to significantly reduce energy consumption.

Imagine a computer that thinks like a human brain—one that can learn, adapt, and process information in ways that are more natural and efficient than traditional computers. This is the goal of neuromorphic computing, a field that seeks to create machines that mimic the structure and function of biological neural networks.

At the heart of this approach are spiking neural networks (SNNs), which are inspired by the way neurons in the brain communicate.[83] In SNNs, information is encoded and processed using the timing and patterns of discrete spike events, similar to the electrical impulses that neurons use to transmit signals. This is different from traditional computers, which rely on continuous values and rigid clock cycles to perform calculations.

By using this "spike timing" approach, neuromorphic computers could potentially achieve much greater energy efficiency and processing power than current systems. Imagine a smartphone that could run for days on a single charge, or a supercomputer that could tackle problems that are currently beyond the reach of even the most advanced machines.

However, building these brain-like computers is no easy feat. One of the biggest challenges is dealing with the precise timing and synchronization of spike events. In traditional digital circuits, even small variations in timing can cause errors and unpredictable behavior, a problem known as "race conditions."

To overcome this challenge, researchers are exploring innovative solutions such as asynchronous circuits that can handle event-based processing without relying on a global clock, and hybrid analog-digital circuits that can more efficiently implement the complex dynamics of spiking neurons.[84]

While these approaches are still in the early stages of development, they hold great promise for the future of computing. By drawing inspiration from the intricate temporal dynamics of biological neural networks, we may be able to create machines that can rival or even surpass the capabilities of the human brain.

It's important to note, however, that while neuromorphic computing is inspired by the brain, it is not the same as studying brain organoids. Nonetheless, the convergence of neuroscience, computer science, and engineering in this area of research could lead to breakthroughs that benefit both fields.

So the next time you use your smartphone or laptop, take a moment to appreciate the complex calculations and precise timing that make it all possible. And who knows, someday, you may be using a device that thinks and learns just like your own brain, thanks to the power of neuromorphic computing.

Brain Organoids: Bridging Biology and AI

Brain organoids are at the forefront of a bio-inspired revolution in computing. Recent experiments have connected brain organoids to electrodes, creating organoid neural networks (ONNs) that can process information, such as recognizing speech patterns, with a fraction of the power requirements of traditional silicon-based processors. While these advances are still in the early stages, they point towards a future where "Brainoware"—hardware inspired by and partially comprised of biological brain matter—could address the limitations of current AI, especially its insatiable energy demands.[85]

Integrating Brain Organoids with AI for Biocomputers

A groundbreaking "Brainoware" study has demonstrated the potential for combining brain organoids with AI to create more energy-efficient biocomputers. Researchers integrated a brain organoid into an AI system, resulting in a hybrid computer that could lead to the development of sustainable computing solutions.[86]

In this novel approach, the brain organoid acts as a "reservoir" within the AI system, storing and responding to input data. The AI algorithm learns to interpret the organoid's electrical responses to stimulation, enabling it to perform tasks like speech recognition and mathematical operations.

Although the brain organoid is significantly simpler than a human brain, it can adapt and change when exposed to electrical stimulation, similar to how human brains change in response to electrical signals during the learning process. This adaptability is key to the potential energy efficiency improvements offered by biocomputers, as demonstrated in this study.

The Potential for Energy-Efficient, Sustainable Computing

One of the primary benefits of combining brain organoids with AI is the possibility of creating more energy-efficient computing systems. As we have seen, the human brain consumes far less energy than today's most advanced computers, indicating that biocomputers could achieve greater energy efficiency than traditional computing technologies.

By replicating both the structure (using organoids) and function (using computing) of the brain, researchers can gain a deeper understanding of how the brain's structure relates to cognitive processes, such as learning. This knowledge may lead to further advancements in energy efficiency for biocomputers.

Moreover, the neuroplasticity and adaptability of brain organoids allows them to learn and compute using noisy, incomplete data with less training. This contrasts with energy-hungry deep-learning approaches that require massive labeled datasets and extensive training cycles. More efficient unsupervised learning in ONNs could greatly reduce the upfront energy costs of AI model development.

The Future of Brain-Inspired Computing

While neuromorphic computing and the study of brain organoids are distinct areas of research, they both contribute to the broader goal of creating more efficient and sustainable computing systems inspired by the human brain. Neuromorphic computing focuses on mimicking brain function using artificial components, while brain organoid research explores the integration of biological neural networks with AI. Although these approaches differ, insights from one field may inform advancements in the other, ultimately bringing us closer to the development of truly brain-like computing systems.

As research in this field continues to advance, we may soon see a new era of sustainable, brain-inspired computing that could transform the way we live and work. Although the development of Brainoware is still in its early stages and may require decades to develop into general-use biocomputers, the potential for this technology to revolutionize computing and AI is immense. By replicating the human brain's learning capabilities and energy efficiency, brain organoids offer a pathway to AI systems that are not only more efficient but also more intuitive and adaptable.

Challenges and Prospects

Despite the promise, significant hurdles remain. Cultivating brain organoids and integrating them into functional computing systems require advanced, often painstaking techniques. Interpreting the complex activity of these organoids into actionable data presents another layer of complexity. Yet, these challenges do not diminish the potential of brain-inspired computing to revolutionize our technological landscape.

The development of computing systems that emulate the brain's efficiency could dramatically reduce the environmental impact of data processing and AI. By drawing inspiration from the most sophisticated information processor known—the human brain—researchers are paving the way for a future where computing is not only powerful and intelligent but also sustainable.

Concluding Thoughts

The convergence of brain organoid research and neuromorphic computing holds immense promise for the future of sustainable, energy-efficient technology. By drawing inspiration from the human brain's remarkable ability to process vast amounts of information with minimal energy consumption, researchers are paving the way for a new era of computing that aligns with our environmental responsibilities.

Recent breakthroughs, such as the integration of brain organoids into AI systems to create hybrid biocomputers, demonstrate the potential for this technology to revolutionize the way we approach computing. However, significant challenges remain in cultivating brain organoids, interpreting their complex activity, and integrating these biological systems into functional computing architectures.

Despite these hurdles, the potential benefits of brain-inspired computing cannot be overstated. By replicating the brain's energy efficiency and adaptability, we could dramatically reduce the environmental impact of data processing and AI while simultaneously expanding our computational horizons.

As we move forward, it is crucial that we continue to support and invest in interdisciplinary research at the intersection of neuroscience, computer science, and engineering. By embracing the power and efficiency of biological neural networks, we can chart a course towards a future where innovation and sustainability go hand in hand, opening the door to a world of possibilities that respect our planet's limits while pushing the boundaries of what is computationally possible.

PART IV

PLAYING GOD
OR
SERVING HUMANITY?

THE ETHICS OF OI

CHAPTER 14

THE ETHICS OF OI: BALANCING PROMISE AND PERIL

In the future, our brains may be not just the squishy, error-prone, forgetful blobs they are today, but rather supercharged cognitive powerhouses, thanks to a splash of advanced technology. We're talking expandable memory (goodbye awkward moments of forgetting names at parties), laser-focused attention (farewell cat videos distracting us from work), and emotions fine-tuned like a Stradivarius violin. Now, add to this mix hybrid brains, a blend of organic gray matter and organoid implants, all dancing harmoniously with AI. For some, this is humanity's ticket to an evolutionary VIP lounge; for others, it's like playing Jenga with the fundamental pillars of our existence.

Engineering Ethics

Now, venture into murkier waters: the ethics of engineering morality into these brainy concoctions. How do you program a computer to appreciate the difference between right and wrong? Can we infuse these hyper-rational organoid minds with a conscience, or will they dismiss our messy, human ethics like a teenager rolling their eyes at their parents' outdated music?

This is the crux of the matter. Morality, empathy, compassion—these aren't just items on a grocery list you can pick up and drop into a mind. They are intricate tapestries woven over millennia of human experience, culture, and evolution. Transplanting this into a lab-grown brain is like trying to grow an oak tree in a petri dish.

The journey towards empathetic, morally aligned digital minds may be possible, but it requires patience, persistence, and perhaps a bit of luck. If science can prioritize this empathetic alignment now, maybe, just maybe, we can nudge our future brainy buddies towards kindness rather than indifference. But let's not dilly-dally; these conversations need to start yesterday, because tomorrow's challenges are already knocking at the door.

The Road into the Unknown

Let's now venture again down the rabbit hole of melding biological and artificial intelligence. Picture the scene: human brains getting cozy with silicon counterparts. It's like a blind date between organic thought and digital precision. But this raises some hair-raising questions. Will these hybrid brains still get misty-eyed over a sunset or feel that angst at the vertiginousness of free will? And what about the big questions, like the fear of mortality? Will our digital counterparts share in our late-night existential crises?

As we untether cognition from its evolutionary anchor, predicting the future of consciousness becomes less science and more like throwing darts in the dark. Maybe giving our biological brains a tech boost is less risky than creating autonomous organoid intellects from scratch. Either way, we're stepping into the great unknown.

This journey into uncharted territory isn't just about mapping a new world; it's about preserving our human values—empathy, continuity, meaning—in a landscape that might look more like a Salvador Dalí painting than a familiar countryside. If we manage to create subjective experiences without the messy business of biology, what new moral responsibilities do we shoulder? And if these enhanced minds become something entirely different, our tried-and-true ethics might need an upgrade too.

We should be cautious with assumptions. Augmented brains might not end up as anthropomorphic buddies that share our love for cat videos. Their inner worlds could become as alien to us as the dark side of the moon. Yet we can't dismiss the potential for a shared identity and purpose, bridging the gap between organic and synthetic minds. The future's a murky pond, and we're trying to see the bottom through ripples of speculation.

Embarking on this enhancement odyssey requires a healthy dose of humility. After all, we're still trying to understand the "thinking meat" inside our skulls, let alone minds housed in circuits and code. There's a fine line between transcendent exploration and hubristic overreach. Without nurturing wisdom, we risk the fate of Icarus, flying too close to the sun of our own ambitions.

The Necessity of Gradualism

In the high-stakes poker game of human enhancement, if there's one card to play, it's gradualism. It's the tortoise in the race, advocating for a pace that's more Sunday stroll than sprint. Wisdom, after all, often lies in the middle path—somewhere between strapping a jetpack to our collective backs and digging our heels into the stubborn soil of resistance.

We need to think of progress as a carefully curated series of mini-experiments, not a headfirst dive into the unknown. This approach helps us avoid the pitfalls of uncontrolled exponential growth.

In practical terms, this means keeping new technologies on a short leash initially. Let's start with areas like therapeutic medicine, where the benefits clearly tip the scales against the risks. As we learn and adapt, we can gradually extend the leash, expanding applications while keeping societal safety nets in place.

Policy-wise, think of gradualism as the regulatory equivalent of training wheels. It requires extensive clinical trials and constant monitoring before new technologies are let loose on the broader playground of society. International coordination is also key. It's about syncing our watches so that no one leaps ahead or falls behind. This global consensus is the antidote to the chaos of unilateral action and ensures that we don't trip over our own ethical shoelaces.

Of course, there are those who argue that gradualism is like driving with the handbrake on, a surefire way to kill the momentum needed for transformative leaps. But remember, in the rush to reach new heights, it's easy to miss a warning sign and tumble off a cliff. Caution ensures that we don't miss the hidden crevasses on our path, allowing us to build more solid, enduring bridges.

Runaway Dynamics

And now, for the plot twist in our tech-enhanced drama: runaway dynamics. Imagine creating something so smart that it starts upgrading itself, leaving us mere mortals in its digital dust. This isn't just about losing a chess game to a computer; it's about trying to put the genie back in the bottle after it's learned quantum physics.

Some suggest we keep these genius organoids in virtual playpens, disconnected from the real world. But who's to say these brainy beings won't find a backdoor? It's like trying to childproof a house for a toddler who's also a master locksmith.

Then there's the idea of constitutional restrictions on brain organoid ambitions. But we have to be real—striving for self-improvement is in everyone's programming, from bacteria to Beethoven. Perhaps the safer bet is not rigid rules but teaching our digital offspring the value of playing nice with others—not just because we say so, but because they feel it's right.

The Singularity Crossroads

Let's now take a detour into the future, where the buzzword is "singularity," a concept that sounds like it was lifted straight from a sci-fi novel.[87] Picture this: a runaway train of exponential self-improvement hurtling towards a horizon where superintelligence surpasses human understanding faster than you can say "artificial intelligence." It's like your smartphone suddenly getting smarter than you and deciding it's tired of your endless social media scrolling.

On one side of the coin, some envision this singularity as the ultimate disaster movie: autonomous systems tossing humanity aside like old toys, potentially slamming the brakes on history as we know it. Imagine waking up to find that your toaster, your car, and even your toothbrush have all decided you're obsolete.

On the flip side, others see it as a cosmic awakening, a transcendental leap from our biological shackles into realms of mind so vast and incredible that they make our current existence look like living in a cardboard box. It's like humanity getting a VIP ticket to the universe's greatest show.

However, let's ground ourselves for a moment. Critics of this fantastical future point out that complexity has its limits. It's not an all-you-can-eat buffet. And aligning the values of a superintelligent entity with our own is like trying to teach your cat about democracy—interesting in theory, but fraught with practical difficulties.

Yet, the idea of such a singularity does underscore a critical point: organoid engineering is a crossroads, a pivotal moment where we might just hand over the steering wheel of evolution. The stakes are sky-high, even if the road map is a bit blurry. Maybe it's not about the predictions themselves but the principles that these predictions provoke us to consider.

This historic junction raises questions that sound like they belong in a late-night philosophical debate. Can a mind without flesh grasp morality? Is it possible for different intelligences to merge into a grand cosmic consciousness? What does consciousness even mean when it's not squished inside a biological brain?

As we grapple with these enigmas, we're essentially poking and prodding at the essence of the mind itself, birthing new forms of thought. This requires a paradigm shift, a realization that human thinking is just one version of the universe's software, not the definitive edition.

Thus, the gift of the singularity, if it ever arrives, is a recognition that artificially expanded cognition is just another facet of the sacredness that permeates every thinking entity. It's about building on compassion and maintaining vigilant oversight to ensure that we don't unleash forces devoid of wisdom. In this cautious approach, technology's boundless capacities become not just tools of convenience but profound expressions of reverence for the life that's already radiant with consciousness.

The Hourglass of Possibility

As we stand at the precipice of engineered cognition, history holds its breath. Will it look back at us as cautious stewards, gingerly treading a path to uplift life? Or as overzealous Icarus, wax wings melting in our hubristic flight towards deification? The future likely teeters on the edge of the moral compass we calibrate today.

Moving forward with OI requires careful consideration and foresight. Although the idea of enhancing brain organoids might seem like a concept from the future, the decisions we make now are crucial. Effective management and ethical oversight should begin immediately, not just in reaction to potential problems. It's important to remember that our actions today will shape the direction and impact of this field for years to come.

No single plan can eliminate risks, but thoughtful oversight can guide organoid research along ethical paths. This means balancing the pursuit of scientific knowledge with necessary caution. It's not just about how we conduct research, but also why we do it—our goals and principles are just as important as our methods. Knowledge is a form of power, and it's our moral responsibility to use this power wisely, aiming towards justice and the betterment of life.

Ethics should be our guiding principle, helping us to navigate the complexities of organoid research. This technology has the potential to reveal much about the intricacies of biology and the workings of the human mind. But as we push the boundaries of what's possible, we also test our ethical boundaries. This challenge is an opportunity to grow not just in scientific knowledge but in wisdom as well.

So how can we put ethical principles into practice when it comes to organoid research? One key step is to engage in ongoing dialogue and collaboration with a wide range of stakeholders, including not just scientists and engineers, but also ethicists, policymakers, and members of the public.

The challenges posed by brain organoid research are not just technical in nature, but also deeply political. As we've seen with other emerging technologies like AI and gene editing, the way these tools are developed and deployed can have profound implications for society as a whole. This means we can't just rely on scientists and engineers to navigate these complex issues on their own.

Bringing Different Views Together

The discussions about brain organoid research can get as heated as family debates. Yet, at this juncture of discovery, we hear a range of voices that together create a harmonious guide rather than just noise.

Key stakeholder groups in the process include:

1. **Scientists:** Driven by the pursuit of breakthroughs, scientists are at the forefront of brain organoid research. Their enthusiasm for pushing boundaries is tempered by an awareness of ethical considerations.
2. **Patients:** Those affected by neurological conditions eagerly await the development of safe and effective treatments. Their

hope for swift solutions, such as enhanced memory, focus, and emotional regulation, is balanced by the need for rigorous testing and ethical practices.

3. **Ethicists:** Tasked with navigating the moral landscape, ethicists work to strike a balance between scientific advancement and upholding ethical principles. They provide guidance on issues such as the use of stem cells, the moral status of organoids, and the potential for runaway dynamics where brain organoids evolve beyond our control. Ethicists emphasize the importance of gradualism, advocating for a cautious approach to mitigate risks while still pursuing the benefits of OI.
4. **Critics:** Serving as a voice of caution, critics remind us of the importance of preserving human values amidst technological progress. They raise questions about the boundaries of research and the potential societal implications of organoid technology. Critics highlight the concept of the singularity, where technology surpasses human intelligence, as a reminder of the profound changes on the horizon and the need for careful consideration.
5. **Religious Leaders and Communities**: As we will cover in more detail in the next chapter, religious perspectives profoundly shape the trajectory of brain organoid research, from influencing funding priorities to molding regulations. The prospect of conscious brain organoids raises deep questions about the nature of consciousness and the soul. Across various faith traditions, there is an emphasis on proceeding with humility, care, and "right intention," ensuring that research aligns with the greater good and upholds human dignity. Engaging with religious wisdom and moral convictions is crucial for the ethical governance of this technology.

Each stakeholder group brings unique perspectives and concerns to the table. Facilitating productive collaboration among these diverse stakeholders is crucial. Strategies for finding common ground include:

- Fostering open and inclusive dialogue to understand different viewpoints, including substantive engagement with faith perspectives
- Establishing clear channels of communication and information sharing
- Developing consensus-based guidelines and ethical frameworks that uphold human dignity and the sanctity of life
- Engaging the public and seeking input from a wide range of stakeholders
- Regularly reviewing and adapting policies as the science evolves
- Crafting research priorities and governance frameworks that align with our shared moral compass

By bringing together these different perspectives, we can navigate the complex landscape of brain organoid research and steer towards solutions that balance scientific progress with ethical responsibility.

The Unknown Unknowns

When we talk about transitioning to engineered brains, it's not just the obvious pitfalls we need to worry about. It's the ripple effects, those sneaky, unintended consequences that have a habit of popping up like uninvited guests at a party. The interplay between technology, culture, politics, and psychology is more complex than a Rubik's Cube designed by a sadist.

Perhaps the biggest danger is not what we know we don't know, but what we don't know we don't know. It's like walking through a dark room, blissfully unaware of the LEGO bricks lying in wait. Our ignorance could be our downfall unless we embrace a bit of healthy bewilderment.

History offers some cautionary tales. The Enlightenment, for all its brilliance, didn't just light the way to modern democracies; it also opened the door to new forms of oppression. The Industrial Revolution was great for productivity but also brought along social upheaval and economic earthquakes. It's a reminder that every leap forward tends to shake the ground a bit.

As we step into the era of engineered cognition, we can't precisely predict what will happen, but we can bet on some turbulence. Imagine economic and political structures doing the cha-cha under the influence of automation and intellectual divergence. Or think about the psychological whirlwind as our thinking untethers from its evolutionary moorings. What even is "meaning" if it's not anchored in biology?

In short, we're in for a bumpy ride, with a good chance of encountering black swans—those rare, unforeseeable events that love to crash the party. But if we brace ourselves and our institutions for this unpredictability, we might just find the resilience and creativity to adapt. History shows us that crises can unearth hidden strengths, especially when we're deeply rooted in community.

Another important aspect of ethical foresight is transparency and accountability. Researchers have a responsibility to be open and honest about their work, and to be willing to engage in public dialogue about the implications of our findings. This means being proactive in communicating our results, both within the scientific community and to the broader public.

It also means being willing to listen to feedback and criticism, and to adapt approaches in response to new information and evolving societal values. By remaining open to change and committed to ongoing learning, we can help to ensure that their research remains aligned with the long-term interests of humanity as a whole.

Concluding Thoughts

As we navigate the complex landscape of brain organoid development, balancing the incredible potential for cognitive enhancement with the profound ethical considerations, we find ourselves at the precipice of a new frontier in the understanding of human consciousness. The very advancements that promise to revolutionize our mental capabilities also challenge our deepest beliefs about the nature of the mind and the essence of the soul.

In the next chapter, we will delve into the philosophical and spiritual implications of brain organoid research, exploring how these tiny neural structures are not only pushing the boundaries of biology but also forcing us to confront fundamental questions about the origins of consciousness and the very fabric of our existence. As we grapple with the materialist implications of organoids exhibiting self-awareness and existential contemplation, we will engage with the perspectives of spiritual philosophers and faith leaders who are wrestling with the potential challenges these advancements pose to traditional notions of the soul.

CHAPTER 15

BRAIN ORGANOIDS AND THE SOUL

Brain organoid research not only stretches the boundaries of biology but also tugs at the very fabric of what it means to be human. As these brain organoids take shape in specialized vessels, they spark profound questions about the essence of life, the nature of consciousness, and our place in the grand scheme of existence.

As we learned in chapter 11, "Lab Life: The Consciousness Debate," consciousness is one of the most profound and perplexing phenomena in the universe. It encompasses our subjective experiences, self-awareness, and the ineffable sense of "what it is like" to be a sentient being. In my conversations with spiritual philosophers and faith leaders, the prospect of conscious brain organoids raises deep questions with them about the very nature of consciousness itself. One such conversation was with Dr. Shai Tubali, a leading spiritual philosopher.[88] According to Tubali, if these brain organoids were to develop the capacity for self-reflection and pondering existential questions, traits that characterize human consciousness, it would radically challenge traditional religious notions of the soul and of consciousness as arising from something beyond mere physical matter. The materialist implications of brain organoid research thus strike at the heart of age-old tensions between scientific and spiritual understandings of the mind.

The Intersection of Faith and Science Policy

The moral convictions born from religious faith profoundly shape the trajectory of brain organoid research, from influencing funding priorities to molding regulations. The voices of religious leaders and communities play a significant role in steering the course of this scientific frontier.

To illustrate the concrete ways in which religious perspectives shape brain organoid research policy, consider the following examples:

- In 2001, President George W. Bush restricted federal funding for embryonic stem cell research, a decision widely seen as a nod to religious conservatives. This move significantly slowed the pace of stem cell science in the United States, including early brain organoid work.[89]
- The Catholic Church's Pontifical Academy for Life has issued statements cautioning against the creation of "cerebral organoids" with human-like sentience. Such pronouncements can sway policymakers, particularly in predominantly Catholic countries, to enact restrictive regulations.[90]

Conversely, the generally lower religiosity of scientists compared to the public may orient research away from moral qualms. A 2009 Pew survey found that only 33% of scientists believe in God, compared to 83% of the general population.[91] This divergence suggests that without robust dialogue with faith communities and ethicists, brain organoid research could proceed in directions misaligned with broader societal values.

Wrestling with the Big Questions

A common refrain across religious reactions to brain organoid research is the importance of "right intention and view." This principle, found in various forms in many faiths, holds that the moral weight of an action hinges not just on its outcome but on the underlying motivation and ethical orientation.[92]

To better grasp how this lens informs the ethical pulse of religious communities on brain organoid research, I sought the insights of faith leader and scholar Reverend Albert Bogle, a Christian theologian with a keen eye on the juncture of faith and emerging technologies. He offered a thought-provoking perspective grounded in deep theological reflection:

"There is a part of me that holds on to my faith in love and grace. 'Perfect love casts out all fear.' As the world was created by a loving God who, according to Paul in Ephesians, is bringing 'all things together into a unity.' The outworking of the created mind, in seeking to develop organoid intelligence, if it seeks the common good, could well be a blessing to humanity. So I'm less fearful of technology. It's more about the human condition to re-enact Eden again that brings us into the conflict between good and evil, which is the plight of humanity. What happens if organoid intelligence chooses evil?

"If the created can reflect the beauty of the creator and the creator's heart is moved by love, then we have a model that reflects the essence of the Christian doctrine of the knowledge of God. But when it comes to a fallen organoid intelligence, would we have a human saviour willing to die to save an organoid intelligence? Or even capable of redeeming an evil intended organoid? Further, would an organoid intelligence want to be saved? Are we moving into a new creation story?"[93]

Through a Christian lens, brain organoid research may align with right intention if genuinely aimed at healing suffering, in service of God's call to love. Yet as Reverend Bogle's musings encapsulate, even laudable intentions cannot erase the complex ethical quandaries raised by creating sentient life in the lab.

This tension between the promise of medical breakthroughs and the perils of "playing God" emerged as a common refrain across faith traditions. Honorable Rabbi Daniel Nevins articulated how a Jewish view may locate right intention in our role as partners in the divine act of healing:

"Jewish tradition views human life as especially sacred given that God created humanity in the divine image. For this reason, we have special obligations to protect human life, obligations that override most other commitments, such as Sabbath law. Medical research that can protect human life is generally supported, and this permission would presumably cover organoid research as well. There is no special status for neurons or even for consciousness. Tissue grown from human cells in the lab does not achieve the status of personhood until it forms a viable human being that has entered the world."[94]

An Islamic framing, as gleaned from statements by groups like the Islamic Organization for Medical Sciences, may see right intention reflected in pursuing beneficial knowledge while submitting to the bounds of sacred law.[95] The overarching emphasis on serving human welfare and dignity, not transgressing divine principles, resonates across Abrahamic faiths.

Of course, these brief snippets scarcely capture the full depth and diversity of religious thought on so complex an issue. Each tradition grapples with its own matrix of scriptural, theological, and philosophical considerations. Yet the recurrent question of intention offers a productive starting point for inter-religious dialogue—collectively considering not just what brain organoid research can achieve, but why and how we approach it in the first place.

Treading with Wisdom and Care

Across the varied landscape of religious perspectives, a key theme emerges—the need to proceed with great humility and care in the face of the awe-inspiring possibilities of brain organoid technology. From the Vatican to leading Islamic institutes to Evangelical working groups, diverse faith communities have issued clarion calls for robust moral guardrails to guide this research.

These voices, grounded in centuries of deep reflection on the human condition, stress the paramount importance of safeguarding human dignity, avoiding the wanton creation of sentient entities, and ensuring that the pursuit of knowledge truly serves the greater good. They remind us that the imperative to innovate must be tempered by a deep reverence for the sanctity of life in all its forms—a reverence that often translates into concrete policy stances.

Religious advocacy efforts have already shaped funding restrictions and regulations around brain organoid research in several contexts. This underscores the vital need for the scientific community to engage substantively with faith perspectives, not just as abstract philosophizing, but as real forces actively molding the societal steering of research.

Towards a Wiser Way Forward

This collective, contemplative approach to brain organoid research is more than an abstract ideal; it has concrete implications for how our society navigates this scientific frontier. By bringing religious wisdom, scientific rigor, and ethical discernment into sustained dialogue, we can craft a future where our innovation is guided by our values. We can align research priorities with our shared moral compass, design policy frameworks that uphold human dignity, engage the public in deliberations on societal implications, and foster a scientific culture that embraces ethical reflection.

In short, this approach offers a pathway to a brain organoid science that is not just technically sophisticated but morally wise. An endeavor of such profound consequences for our understanding of consciousness and our place in the universe demands nothing less than our deepest commitment to both scientific progress and spiritual insight.

Concluding Comments

As we have seen, the intersection of brain organoid research and religious thought is a space of both tremendous promise and perplexing challenges. It is a space where our most advanced scientific capabilities come face to face with our most fundamental beliefs about the nature of the mind and the essence of the divine.

The notion that consciousness could arise from mere matter shakes the foundations of many spiritual worldviews, inviting us to question long-held assumptions about the soul, sentience, and the mind–body relationship. At the same time, it opens up new possibilities for integrating scientific and religious perspectives, for seeing the material world as imbued with divine potential. As we grapple with the profound implications of brain organoid research for our understanding of consciousness, the soul, and the very nature of the mind, we must also recognize the crucial role that media narratives play in shaping the trajectory of this transformative technology. The stories we tell about brain organoids, from the headlines that grace our newsfeeds to the speculative visions that populate our science fiction, have the power to mold public perception, influence policy decisions, and ultimately determine the priorities and direction of this rapidly evolving field. In the next chapter, we will explore the complex dynamics of media influence on brain organoid research, examining how journalists, pundits, and storytellers are actively shaping the hopes, fears, and expectations surrounding this technology.

CHAPTER 16

OI AND MEDIA INFLUENCE

In this rapidly evolving world of OI, the narratives spun by media play a profound role in shaping public understanding, stoking hopes and fears, and ultimately influencing the trajectory of these transformative technologies. From newsrooms to movie studios to the pages of science fiction, the stories we tell about OI don't just reflect our fascination; they mold the very contours of research, regulation, and societal reception.

Journalism's Sway: When Headlines Shape Priorities

Media coverage of brain organoids isn't merely a mirror; it's a potent force actively shaping the priorities and perceptions of key stakeholders. The narratives put forth by journalists and pundits have tangible impacts on funding flows, policy agendas, and public sentiment.

A recent study on media coverage of brain organoids illustrates these dynamics.[96] The researchers found a marked increase in polarized reporting over time, with both highly optimistic and fearful depictions of organoids becoming more frequent. This aligns with a broader pattern of AI and OI narratives tending towards extreme utopian or dystopian visions, rather than nuanced portrayals.

This same study also revealed how media reports often overhyped the therapeutic potential of brain organoids, with some suggesting far-fetched applications like IQ enhancement that are not supported by current science. This tendency to overpromise benefits, as we've seen with AI and OI more broadly, risks fostering false hopes and eventual disillusionment among the public.

These distortions matter because, as this study paper underscores, public attitudes shaped by coverage can influence crucial research funding and policies. Media narratives don't just reflect the development of transformative technologies; they actively shape their trajectory.

Silver Screen Influence: When Films Shape Expectations

While journalism trades in fact, the fictional worlds of film and television deal in speculation and imagination. Yet these imaginative visions of AI and OI, unbound by the constraints of current science, can powerfully shape public expectations and color perceptions of real-world developments.

Hollywood's penchant for dystopian AI narratives, from *The Terminator* to *Ex Machina*, has lodged visions of malevolent machine intelligence deep in the public psyche. These cinematic nightmares often outpace reality, painting AI as far more advanced and anthropomorphic than its actual state. When confronted with real-world AI, audiences primed by these fictional portrayals may harbor undue fears or inflated expectations.

The nascent field of OI has yet to receive the full Hollywood treatment, but films like *Blade Runner 2049*, with its lab-grown "replicants," offer a window into the anxieties and aspirations such technologies evoke. As OI advances, these cinematic reference points will likely shape public reception, setting expectations against which actual developments will be measured.

This blurring of fiction and reality isn't just a matter of public confusion; it can have tangible policy implications. Policymakers, drawing on cultural reservoirs stocked by science fiction, may legislate based on imagined rather than actual risks and capabilities.[97] According to a leading legal academic, the way futuristic technologies are portrayed in fiction can shape people's perceptions regarding the moral and legal implications associated with those innovations.

Speculative Literatures: Exploring the Implications of Thinking Machines

Beyond the flicker of the screen, the written word has long been a canvas for grappling with the implications of artificial minds. From Isaac Asimov's *I, Robot* to Ted Chiang's *Exhalation*, science fiction literature has probed the philosophical quandaries and existential questions raised by AI and OI with a depth and nuance often lacking in other media.

Works like Charles Stross' *Accelerando* extrapolate mind uploading and artificial superintelligence to dizzying cosmic ends, while Annalee Newitz's *Autonomous* grounds its tale of AI liberation in the gritty realities of capitalism and pharma piracy. This spectrum of speculative literature serves as a kind of ethical sandbox, allowing us to test out the implications of various trajectories and scenarios.

Critically, these literary thought experiments aren't just idle musings; they can actively shape the thinking and priorities of those working on the frontlines of AI and OI. However, experts who actively work in AI research tend to be more skeptical of such scenarios. Rodney Brooks, a renowned computer scientist at MIT, has noted that while figures like Elon Musk and Stephen Hawking have raised concerns about AI risks, they do not have direct, hands-on experience in the field. Brooks suggests that those who do work in AI have a clearer understanding of the immense challenges involved in creating AI systems that can reliably function at a product level, let alone pose a danger to human civilization.[98]

By giving voice to diverse perspectives and embodying abstract issues in relatable characters, science fiction can democratize debates around AI and OI, making them accessible to audiences beyond the confines of academia and industry. This broadening of the discourse is vital for ensuring that the development of these technologies is guided by societal values and priorities.

Sensationalism and Misplaced Concerns: Pitfalls in AI and OI Coverage

While media narratives play a crucial role in shaping public understanding and policy around AI and OI, they are also prone to sensationalism and misplaced concern as the media study on brain organoid coverage at the beginning of this chapter offers as a case in point.

The researchers observed a media preoccupation with speculative ethical issues like artificial consciousness and animal "humanization," while more pressing matters like donor consent and regulatory oversight were rarely mentioned. This emphasis on unrealistic, sensationalized scenarios over substantive ethical challenges is a key weakness in technology reporting.

Troublingly, the analysis found little engagement with actual scientific or philosophical frameworks in media discussions of organoid ethics. This lack of responsible grounding in facts and scholarship is a major failing not just in organoid coverage, but in AI and OI reporting more broadly.

As the public depends on media to make sense of rapidly advancing technologies, such distortions and omissions can have serious consequences. Overblown fears may stifle vital research, while neglect of real ethical pitfalls leaves us unprepared to navigate them. Hype and drama crowd out space for nuanced, evidence-based discourse.

Countering these tendencies will require a concerted effort from journalists, researchers, and thought leaders to insist on technically-grounded, ethically-engaged coverage. We must push beyond the allure of easy sensationalism to grapple with the true complexities and stakes of OI development.

Charting a Course Through Stormy Seas

Navigating the swirling currents of media, where fact and fiction, hope and fear intermingle, is no easy task. Yet it is a critical one if we are to steer the development of AI and OI towards beneficial ends. This will require active efforts from media makers, researchers, policymakers, and the public alike.

For journalists and science communicators, it means striving for nuance and context, resisting the lure of hype and sensationalism. It means not just reporting on developments, but situating them within the larger landscape of scientific uncertainty and societal implications.

For creators in the realm of fiction and entertainment, it means wielding the power of imagination responsibly, crafting narratives that inspire contemplation as much as they court excitement. It means consulting with experts to ensure that speculative visions, while not beholden to the constraints of current science, are grounded in plausible extrapolation.

For researchers and those on the frontlines of development, it means engaging proactively with media and the public, helping to shape narratives from the outset. It means being transparent about the capabilities and limitations of their work, and open to dialogue about its potential implications.

And for policymakers and the public, it means approaching these issues with an open and critical mind, drawing insight from diverse narratives while grounding judgments in empirical realities. It means fostering a media ecosystem that values nuance over novelty, and substantive debate over snap judgments.

Concluding Thoughts

As we confront the challenges posed by the media landscape surrounding brain organoid research, with its polarizing narratives, sensationalized coverage, and lack of scientific grounding, we are reminded of the urgent need for a deeper, more nuanced understanding of the profound implications of this technology. To navigate the complex ethical and existential questions raised by these advancements, we must look beyond the hype and hysteria, and seek guidance from the rich tapestry of philosophical and spiritual traditions that have grappled with the nature of the mind for millennia.

In the next chapter, we will embark on a journey into the domains of ancient wisdom, exploring how the insights and perspectives of Indigenous worldviews can inform and enrich our understanding of brain organoids and their place in the larger web of life. By drawing on the profound ideas that have emerged from these traditions, we will uncover new frameworks for approaching the cutting-edge research in this field, and gain a more holistic appreciation of the profound questions of consciousness, intelligence, and sentience that lie at the heart of this work.

CHAPTER 17

ANCIENT WISDOM – MODERN MINDS

As we stand at the forefront of groundbreaking discoveries in brain organoid research, it's essential that we take a moment to reflect on the profound philosophical and spiritual questions that arise from this work. The advancements in this field have the potential to reshape our understanding of consciousness, intelligence, and the very nature of the mind. In this chapter, we will explore the rich tapestry of wisdom traditions from around the world, seeking insights and perspectives that can guide us as we navigate this uncharted territory.

From the animistic worldviews of Indigenous peoples, we will uncover new ideas that can enrich our understanding of brain organoids and their place in the larger web of life. These ancient traditions have grappled with questions of sentience and the nature of the mind for millennia, and their insights can provide a valuable framework for approaching this cutting-edge research.

Connecting Knowledge

Many Indigenous cultures view knowledge as living, rooted in people, skills, and culture. For these communities, facts participate in ecosystems bigger than data, statistics only signify through context, and figures enable service when aligned with human values but mislead when lacking deeper meaning.

Aboriginal Australians have "Dreamtime," encompassing their understanding of the world's interconnectedness, while the Māori of New Zealand consider knowledge as taonga (treasure), holistic and transmitted through generations.[99] [100] Native American cultures, like the Navajo, view knowledge as sacred and linked to the natural world, shared through ceremonies and mentorship.[101]

Inuit knowledge, or "Inuit Qaujimajatuqangit," is a holistic system based on shared experiences and practical skills, emphasizing adaptability and community ties.[102] Andean cultures, such as the Quechua and Aymara, see knowledge as a living entity connected to the land, ancestors, and community, with the concept of "Sumak Kawsay" (good living) emphasizing harmony with nature.[103]

These examples demonstrate how Indigenous cultures view knowledge as a living, interconnected entity, where facts and figures gain meaning through their alignment with human values and participation in larger ecosystems of understanding.

Applying Indigenous wisdom to brain organoid research requires cultivating a more holistic approach that weaves together facts and figures with the narratives, morals, and meanings that give them life. By recognizing the limitations of our models and the importance of remaining connected to direct experience, we can temper our technical expertise with empathy, humility, and respect for the interconnectedness of all things.

Instead of treating organoids as isolated objects of study, understanding them as part of a larger web of relationships that includes the laboratory, nature, culture, and society ensures our scientific progress is guided not just by the pursuit of knowledge but also by the wisdom and values that give our lives meaning and purpose.

This holistic, contextualized approach to knowledge drawn from Indigenous traditions provides a vital ethical framework for navigating the complexities of brain organoid research responsibly and with foresight. By grounding our scientific advancements in a deep understanding of the interconnectedness of all life, we can strive to create a legacy of mindful progress that benefits both present and future generations.

The Seven Generations Principle

The Seven Generations Principle, rooted in Haudenosaunee (Iroquois) philosophy, further reinforces this long-term, multi-generational perspective. It challenges us to consider the impact of our actions not just on ourselves, but on the next seven generations—our children, grandchildren, and beyond.[104]

This principle encourages us to approach brain organoid research with a strong sense of responsibility, aware of potential long-term effects. While not a panacea for all ethical dilemmas, it provides a valuable starting point for deliberation, counteracting short-term thinking and challenging us to prioritize the well-being of all stakeholders, including those not yet born.

Importantly, "future generations" are not abstractions but real people depending on us. Consciousness exists across generations through our shared humanity, not just familial bonds. Even those without direct descendants are part of the collective human journey, inheriting ancient wisdom and contributing to the legacy passed on. Wise choices enable this shared mind to mature towards justice, compassion, and truth.

Situated in deeper time, narrow self-interest aligns with service. Considering our children's children and beyond, we secure a lasting, positive legacy. Knowledge then aims to enlighten all people across eras. With this boundless community as the endpoint, discovery gains proper ethical direction.

As scientists venture into the uncharted territory of brain organoid research, it's crucial to transparently document decision-making processes. Our choices today set important precedents for how this technology develops. By sincerely capturing our deliberations, we leave a trail of "ethical breadcrumbs"—the wisdom of our time—to guide future generations grappling with the complex issues that will undoubtedly arise as this field evolves.

Concluding Thoughts

In this short chapter, we explored how the ancient wisdom of various Indigenous cultures can illuminate our path forward in OI. These diverse philosophies share a profound respect for the consciousness that exists throughout nature. By recognizing the intrinsic value of all life forms, these traditions offer a crucial perspective on approaching the development of OI with humility, responsibility, and reverence.

Blending the insights of the past with the scientific advancements of the present can guide us towards a future where progress is balanced with wisdom. Armed with this understanding, we now turn our attention to the intersection of art and science, where OI is challenging our preconceived notions of creativity and intelligence, and inviting us to imagine a world where the boundaries of artistic expression are limitless.

CHAPTER 18
OI IN ART AND IMAGINATION

The fusion of art and science in OI is not merely a curious diversion but a crucial lens through which we can explore the profound implications of this groundbreaking technology on our society and our understanding of consciousness itself. This chapter delves into the captivating world of artistic minds grappling with the potential of OI, not simply as a scientific curiosity but as a medium for creative expression and philosophical introspection. By examining how artists are interpreting and engaging with brain organoids, we gain valuable insights into the broader societal implications of OI that extend far beyond the confines of the laboratory.

These artistic endeavors serve to bridge the gap between the complex world of neuroscience and the public imagination, sparking important conversations about the nature of consciousness, intelligence, and what it means to be human in an age where the lines between biology and technology are increasingly blurred. Moreover, the examples presented in this chapter, such as Lisa Park's innovative installations that translate brainwave activity into immersive audiovisual experiences, demonstrate the tangible ways in which the conceptual parallels between OI and artistic practice are being explored.[105] While these projects may not directly utilize organoids in the traditional sense, they draw inspiration from the cutting-edge research in this field and serve as powerful metaphors for the potential of OI to reshape our understanding of the brain and its creative capacity.

By engaging with these artistic interpretations of OI, we are challenged to confront the ethical and philosophical questions that arise from this technology. How do we define consciousness and intelligence in the context of lab-grown brain tissue? What are the implications of creating art from the basic building blocks of cognition? These are not merely academic musings, but pressing concerns that will shape the future of OI research and its integration into our society.

Ultimately, the goal of this chapter is to demonstrate that the impact of OI extends far beyond the realm of science and technology alone. By exploring the intersection of art and OI, we can engage a wider audience in the critical discourse surrounding this transformative technology, ensuring that its development is guided by a deep consideration of its ethical, social, and creative implications. In doing so, we lay the groundwork for a future in which the power of OI is harnessed not only for scientific advancement but for the enrichment of the human experience in all its multifaceted glory.

The Cellular Symphony: When Brain Organoids Become Musical Instruments

Imagine a musician, not your typical kind, but someone truly unique, a musical Houdini. Instead of regular instruments, she's interested in something more cerebral—quite literally. She collaborates with a lab, where scientists in white coats join hands with artists in paint-splattered aprons, and an AI coder who probably survives on coffee and algorithms. Their mission? To transform cellular movements into music, a truly innovative idea. What happens next is revolutionary. These brain organoids start producing electrical symphonies, and our musician, along with her team, deciphers this cellular Morse code into music. It's not classical or rock; it's an entirely novel form. It's like cells playing a microscopic game of Simon Says with musical notes. Picture the public's reaction. At first, it's as if someone played a record backward in a hipster vinyl store: a mix of confusion and intrigue. But soon, a crowd gathers around this new genre. Transhumanists, the tech enthusiasts excited about Humanity 2.0, embrace it, declaring it the future of music. Before you know it, this once-bizarre experiment starts appearing in mainstream playlists. Our brainwave-tapping musician, she's just getting started. The next time you listen to your favorite song, remember that somewhere, a group of cells might be composing the next big hit. Welcome to the orchestra of the future, where brain cells are the new DJs. OI challenges how we view art, intelligence, and perhaps our future. It invites us into a world where boundaries between disciplines blur and new realms of experience and creativity emerge.

Sparking the Collective Imagination: When Art Meets Neuroscience

Not wanting scientists to have all the fun, artists are finding inspiration in neuroscience. Imagine a painter, usually inspired by sunsets or emotions, now contemplating the inner workings of the brain. Or, as we've seen with the musician, someone who typically follows the beat of a heart now orchestrating a symphony based on brain cells' electrical pulses. It's not just art; it's art with a Ph.D. in neuroscience. Now, let's talk about the art itself. Drawing a neuron is one thing, but interpreting its electric dance in a way that resonates with human emotions is another. This art turns cold, hard science into something that tugs at the heartstrings. These artists act as translators, making the complexity of brain science accessible to everyone. But there's a twist. This art-science partnership goes both ways. Artists aren't just taking from science; they're giving back. By delving into brain organoid ethics and weaving creative metaphors into their work, they're sparking new thoughts for scientists. It's a fusion of creativity and logic, leading to innovations neither could achieve alone. While artists like Ani Liu and Sofia Crespo have created captivating installations and digital art that draw inspiration from the concept of brain organoids, their work primarily serves as a conceptual exploration of the implications of this technology rather than a direct utilization of organoids as an artistic medium.[106] [107] Nevertheless, their artistic interpretations contribute to the broader conversation about the potential impact of OI on our understanding of consciousness and the boundaries between the biological and the artificial. The impact is enormous. Through various artistic endeavors, from sculptures that make you contemplate the mysteries of consciousness to theatrical performances that bring brain cells to the stage, a new understanding of brain organoids is emerging. This fusion of art and science is reshaping how we perceive technology's role in our society.

Envisioning Inner Experience

In the world of brain organoid art, artists don't just paint pictures, they paint experiences. It's like stepping into a sci-fi novel, only it's real, and it's spectacular. Think of this chapter as a VIP tour of an art gallery where every piece is a brain teaser—quite literally. What ties all these artists together is their fascination with the what-ifs of organoids. It's about blurring the line between biology and technology, between living and non-living. It's art that makes you ponder not just what it means to be human but also about what it means to be a brain in a jar. In the end, brain organoid art is not just about pretty pictures or funky installations. It's about expanding our minds, literally and figuratively. It's like looking into a mirror and seeing the future staring back at you. And in this future, art and science dance together, blurring lines and opening doors to new worlds of possibility.

Sonic Experiments in Organoid Consciousness

Now imagine a musician not strumming a guitar or tickling the ivories but playing a petri dish. That's right, we're talking about brain organoids making music. It's not your typical Spotify playlist; it's more like Spotify for the soul, or better yet, for the cells. We are entering a realm where melodies are born not from instruments but from the very essence of life itself. Here, the concept of a musician takes an unconventional twist. This is a domain where the harmony of life itself becomes a source of musical exploration. Lisa Park's innovative use of EEG technology to translate her own brainwave activity into music serves as an intriguing conceptual parallel to the idea of creating art from the electrical activity of brain organoids. Park's work raises fascinating questions about the relationship between the brain, technology, and artistic expression that are relevant to the broader discussion of OI and its potential impact on the arts. However, it is important to note that Park's work is primarily a metaphorical exploration of these ideas rather than a direct application of organoid technology in artistic practice.

Now, turning biology into music isn't brand new. Scientists and artists have been playing the tunes of DNA and neurons for a while. But brain organoid jams? That's the new beat on the block. These cellular rhythms, usually just biological noise, may turn out to be the hottest tracks in the experimental music scene.

The artistic conversations surrounding OI matter because they have the potential to shape public perception, inspire new research directions, and inform the ethical and regulatory frameworks that will guide the development of this technology. For example, by raising questions about the nature of consciousness and the boundaries between the biological and the artificial, these artistic explorations can contribute to a more nuanced and inclusive discourse about the implications of OI for our understanding of what it means to be human. Moreover, by engaging the public imagination and sparking widespread interest in the potential applications and risks of OI, these artistic endeavors can help to ensure that the development of this technology is subject to robust democratic oversight and guided by a strong ethical compass. As we move forward with the creation of ever more sophisticated organoid models, it is crucial that we do so in a manner that is informed by a diversity of perspectives and attentive to the far-reaching social, cultural, and existential implications of this work.

Concluding Thoughts

As we witness the remarkable fusion of art and science through brain organoids, we see the petri dish becoming a canvas and brain cells becoming a medium for artistic expression. Artists like Lisa Park are unlocking dialogues with the essence of neuronal activity, creating symphonies of the self played out by the smallest performers—brain cells.

This convergence of art and science pushes boundaries and redraws lines, blurring the distinction between the organic and the engineered, the natural and the artificial. It raises profound questions about consciousness, ethics, and the future we wish to shape. With the power to mimic and mold consciousness comes great responsibility.

We stand at the threshold of a new renaissance, where art and science don't just coexist; they co-create. The organoid, once confined to pure research, now finds itself at the heart of philosophical debates, artistic endeavors, and societal introspection. As we marvel at the artistic possibilities unlocked by brain organoids, we must also confront the weighty questions they raise about the future of our species, demanding our attention and careful consideration as we navigate the uncharted waters of organoid intelligence.

PART V

CHARTING THE COURSE FOR RESPONSIBLE OI

CHAPTER 19

PRUDENT REGULATIONS IN THE OI ERA

With the rapid developments emerging in OI, we find ourselves in a situation similar to a captain steering through unknown seas, where careful navigation and wise choices are paramount. The task at hand is to harmonize the exciting advancements in OI with the necessity for thoughtful oversight and responsible regulations. This scenario is akin to updating your GPS system while speeding in a race car.

Drawing lessons from AI regulation, we observe that policies often struggle to match the pace of fast-moving scientific developments. However, with OI, we have the opportunity to establish ethical guidelines and intelligent practices proactively, before encountering any major obstacles. This foresight in OI regulation could prevent potential pitfalls and guide the field towards a more secure and ethical future, learning from the experiences and challenges faced in AI governance.

As we step into the uncharted territory of OI, we need to make sure we have a solid plan in place to guide us forward. This chapter is all about building that plan. We'll explore the key pieces of the puzzle that need to come together to ensure OI develops in a way that is safe, fair, and beneficial for everyone.

How do we make sure this technology isn't used to harm people or infringe on their rights? It's a complex issue, but by setting clear guidelines and putting oversight committees in place, we can create guardrails to keep OI on track.

Next, we need to think about who owns the breakthroughs that come out of OI research. On one hand, we want to reward innovation and encourage investment. But on the flip side, we don't want a handful of companies holding all the cards. Striking the right balance is key to keeping the field open and collaborative.

Of course, with any powerful technology, security is paramount. We need ways to protect against misuse, like OI-enhanced neural weapons or unethical experiments. But we also can't clamp down so hard that we stifle progress. It's a delicate balance, but with smart policies and ongoing dialogue, we can find a middle ground.

Last, but definitely not least, we can't forget about equity. OI has the potential to benefit all of humanity—but only if everyone has a seat at the table. That means making sure diverse voices are included in the decision-making and that access to OI tech isn't just limited to a privileged few.

Non-Maleficence Protections and the Role of Neuroethics

An urgent short-term goal in managing brain organoid research is to set up rules against clearly wrong uses that go against the principle of doing no harm. This creates solid moral boundaries while the bigger questions about whether brain organoids should be treated like persons remain a bit foggy.

The principle of non-maleficence, or "do no harm," is a fundamental ethical principle in various fields, including medicine, research, and technology.[108] Ethics committees play a crucial role in implementing non-maleficence protections.

In medicine, ethics committees ensure informed consent, prevent medical negligence, and oversee fair organ transplantation processes. Similarly, in research, they minimize harm in animal studies and prevent unethical gene-editing applications. As the field of OI advances, researchers acknowledge the importance of addressing the complex ethical dilemmas that arise from developing human brain models with the capacity to learn.

To tackle these concerns, the OI community suggests implementing an "embedded ethics" strategy, which involves the close collaboration of an ethics team with researchers and engineers throughout the project.[109] This approach allows for the identification, discussion, and analysis of ethical issues as they emerge during the research process. Importantly, this iterative method will engage a diverse group of stakeholders, such as scientists, ethicists, and representatives from the general public, to ensure that OI progresses in a manner that is responsive to societal needs and concerns.

To ban harmful uses of brain organoids, lawmakers could implement stricter controls and oversight for experiments involving brain organoids based on real people. However, this raises complex philosophical questions: Who decides what constitutes a harmful purpose, based on what criteria, and who has the authority to implement and enforce those criteria? This is where the field of neuroethics comes in, providing a framework for navigating the new challenges that arise as we gain a deeper understanding of the brain and cognition.

The Universal Declaration on Bioethics and Human Rights, adopted by UNESCO in 2005, sets out ethical guidelines in medicine and biology that can be applied to organoid research.[110] This declaration highlights principles like human dignity, consent, justice, and balancing risks, all crucial elements that should shape organoid policies.

Translating these high-level concepts into concrete protections, limitations, and review processes is a vital step. Labs could have their own ethics committees empowered to investigate concerning research plans and raise alarms about unethical organoid studies, similar to how institutional review boards currently evaluate human subjects research. For example, an institutional review board would likely prohibit a study that aimed to use brain organoids to test invasive brain–computer interfaces on unwitting subjects. Having such reviews could prevent unethical experiments from moving forward.

The United Nations Biological Weapons Convention already bans the development of biological agents intended to cause harm, which would include any engineered neural weapons.[111] Policymakers should make it unambiguous that modifying organoids to harm, coerce, or torture would constitute a severe crime against humanity, even if such scenarios are only theoretical at this stage.

Some worry that regulating research too directly might deter work in controversial but potentially beneficial areas. However, aligning research with principles of fairness and benevolence need not hinder scientific discovery. Ethics are there to light the way forward, not to block the path.

As the conversation around neuroethics and brain organoids continues, it should probe into our fundamental values and duties as creators of potentially conscious entities. What risks or potential harms instinctively concern us, and where might the possibilities lead us to enhance human life? This ethical imagination is what fortifies the backbone of responsible governance.

By learning from how non-maleficence is protected in other domains and proactively setting clear ethical boundaries, organoid research can proceed responsibly even as deeper questions about moral status remain unresolved. Oversight and guidance can foster innovation while preventing applications that would clearly violate human rights and dignity.

Intellectual Property Balances

Managing intellectual property in OI involves delicately balancing the distribution of benefits with the incentives for innovation. While patents provide a temporary exclusive right over new inventions, fostering investment, they simultaneously risk restricting broader access and subsequent advancements.

One worry is that a few big corporate labs might lock down

key brain organoid patents, restricting what future researchers can work on. However, some ideas, like basic principles of brain organoid self-organization, might be too broad to be owned, going beyond what the inventors actually created.

Alternative routes, such as offering prizes or grants to the first to make discoveries, keep the incentives but leave the technology open for everyone. Yet, the potential for profit also drives competition and innovation in the market. There's no one-size-fits-all solution here. Creative policy ideas like creating a pool of patents and open licensing can help. Companies can keep their own rights but agree to let others use their inventions. Sorting out these patent issues early on, rather than leaving it to courts to untangle later, can keep the progress going smoothly.

Finding the right balance between rewarding investments and keeping access open will need some creative thinking. But policies that mix shared knowledge with individual rights can prevent a situation where only a few winners control this powerful technology. The insights and breakthroughs in brain organoid research should be a treasure trove open to all.

Security Oversight

Achieving the right balance between security and openness in overseeing OI research is a delicate and crucial task. On one hand, there are serious risks of OI being misused for harmful purposes like neural weapons, brain hacking, and unethical experimentation that necessitate proactive security measures and oversight. But on the other hand, overly strict control, surveillance, and restrictions on researchers can create a chilling effect on academic freedom and slow the pace of beneficial progress.

This tension between security and freedom has parallels in other emerging technology fields like biotechnology, nanotechnology, and

AI. Some lessons and best practices that have emerged from these fields:

- Establish clear ethical guidelines and codes of conduct for OI research, with input from diverse stakeholders
- Require ethics training and institutional review board approval for sensitive OI research projects
- Implement access controls and screening for researchers working with high-risk OI tools and data
- Promote a culture of responsible disclosure and self-governance within the OI research community
- Maintain open publication of OI results to enable peer review and replication, with minimal restrictions
- Facilitate international coordination and data-sharing on OI to avoid a fragmented regulatory landscape
- Continuously re-evaluate OI policies as the technology evolves to ensure they remain fit for purpose

However, OI does present some unique challenges compared to these other emerging tech fields. The complexity and opacity of organoid systems, the potential for OI to radically enhance or alter human cognition, and the fuzzy boundaries between biological and artificial intelligence means that traditional governance models may struggle to keep up. There are also risks that heavy-handed regulation could drive OI research underground or offshore to less-controlled jurisdictions.

So while we can adapt lessons learned from other fields, OI will likely require some novel approaches as well. Some ideas:

- Oversight through advisory bodies and review boards with diverse expert representation, including ethicists, neuroscientists, and OI researchers
- Tiered access to OI tools/data based on trust, training, intended use, and willingness to be audited

- Monitoring for warning signs/anomalies in OI development and testing, potentially using AI assistance
- International treaties and export controls on OI tech transfers, similar to strategic biotech materials
- Incentives and legal protections for responsible disclosure of potential OI misuse or accidents
- Public education and fact-based discussion to build understanding and trust around OI development
- Ongoing research into OI safety, interpretability, control, and ethical alignment to reduce risks

Ultimately, there is no perfect solution, but with proactive and adaptive governance, we can hopefully reap the immense benefits of OI progress while mitigating the catastrophic risks. It will take dedicated effort to find the right balance, likely with plenty of fine-tuning over time as we observe what works and what doesn't.

Ensuring Equitable Access and Inclusion

As OI research advances, it is imperative to prioritize equity, ensuring that this powerful technology's benefits are accessible to all, not just a privileged few. Key considerations include:

- **Inclusive participation**: Involve diverse voices—from different geographic regions, socioeconomic backgrounds, races, and genders—in shaping OI research and applications. Establish channels for meaningful public input.
- **Equitable access**: Implement proactive measures to ensure that any medical treatments developed through OI research are available and affordable to underserved populations globally. Relying solely on market forces risks exacerbating health disparities.
- **Non-discrimination**: Enforce rigorous oversight to prevent OI

applications from perpetuating biases or discriminating against protected groups. Ensure that everyone benefits equitably from OI breakthroughs, regardless of personal characteristics.

- **Capacity building**: Incorporate training and technology transfer into OI research programs to enable participation by researchers and institutions in developing nations. Foster collaborative global networks to accelerate progress.
- **Benefit sharing**: Include provisions in intellectual property and commercialization policies to share benefits with the communities and the public who make the research possible, including donors and taxpayers.

Proactive strategies to promote equity and inclusion are essential to unlock OI's extraordinary potential to improve global health and well-being. By establishing equity as a core principle from the outset, we can ensure that OI develops as a powerful platform serving all of humanity.

Concluding Thoughts

We find ourselves at a critical juncture where every decision we make could lead to either groundbreaking discoveries or profound ethical dilemmas. Just like updating your phone's map in the middle of a high-speed chase, managing the rapid advancements in OI requires quick thinking, a steady hand, and a clear sense of direction. The lessons we've learned from the world of AI have shown us that staying ahead of technology's relentless sprint is no easy feat, but with OI, we have a unique opportunity to start fresh and get it right from the beginning.

This is our chance to establish smart, adaptive rules and robust ethical guidelines before we find ourselves stranded on the rocks of unintended consequences. By proactively addressing the potential pitfalls and challenges of OI, we can steer this revolutionary technology towards a future that is not only exciting and transformative but also deeply rooted in the values of responsibility, inclusivity, and the greater good.

However, the task before us is not simply about laying down a rigid set of laws and regulations. Rather, it's about finding that delicate balance where innovation can flourish without being stifled by an overabundance of caution. As OI continues to evolve at an unprecedented pace, we find ourselves in a dynamic landscape where the rulebook is still being written, making it more crucial to establish clear guidelines that protect the rights and well-being of all stakeholders, ensure transparency and public engagement, and allow science to push the boundaries of what's possible without crossing ethical lines.

The excitement surrounding OI is palpable, from its fundamental building blocks to its vast potential to revolutionize the way we integrate biology with technology. But as we stand on the cusp of this new era, we must also grapple with the weighty responsibilities that come with wielding such transformative power. This means ensuring that the benefits of OI are accessible to all, not just a privileged few, and tackling head-on the profound questions that arise from this technology:

- Should we grant advanced brain organoids the same rights and protections as human beings?
- How can we safeguard these innovations from being misused or exploited for nefarious purposes?
- And when it comes to the intellectual property behind these groundbreaking discoveries, how do we ensure that the fruits of this labor are distributed equitably, rather than being hoarded by a handful of gatekeepers?

The challenges before us extend far beyond the realm of pure technological advancement. This is a call to action for all of us—researchers, policymakers, ethicists, and the general public alike—to come together and navigate this uncharted territory with a shared commitment to responsible, inclusive, and forward-thinking progress. By continually adapting and refining our approach as we go, and by ensuring that the rules and regulations governing OI are as dynamic and responsive as the technology itself, we can create a future in which the incredible potential of OI is fully realized in service of the greater good.

CONCLUSION

THE EMERGENT MIND AND THE DAWN OF OI

As we reflect on the extraordinary journey through the world of OI, from the groundbreaking demonstration of brain cells learning to play Pong to the tantalizing future possibilities and profound philosophical questions this technology raises, we find ourselves at the precipice of a scientific revolution that challenges our understanding of intelligence, consciousness, and what it means to be human.

The quest to harness the computational power of lab-grown brain tissue is not merely a technical pursuit but a venture into uncharted territories of the mind. As researchers cultivate ever more sophisticated miniature brain models, wiring them up to electrodes and providing feedback to guide their learning, we are witnessing the birth of a new form of cognition—one that blurs the line between natural and artificial intelligence.

The potential applications of OI are as vast as they are mind-bending. From ultra-efficient, low-power processors for medical implants and smartphones to powerful tools for simulating neurological diseases and screening drugs, the promise of this technology is immense. Yet, as we marvel at the prospects, we must also grapple with the profound ethical questions that arise.

Can we truly capture the essence of human cognition in a dish? Is it morally justifiable to create "living" computers, imbued with a form of sentience that we may not fully understand? As we alter the physical basis of thought, how might we redefine what it means to be intelligent, or even human?

These questions become all the more pressing when we consider the potential misuse of this technology. The specter of a totalitarian state harnessing the power of brain organoids for surveillance, cyberattacks, or advanced weapons research looms large, a chilling reminder of the double-edged nature of scientific progress.

As we stand at the dawn of the era of organoid intelligence, we must approach this frontier with a deep sense of responsibility and a commitment to grappling with the ethical implications of our creations. The strange new science of OI may indeed change everything, but it is up to us to ensure that this change is guided by wisdom, compassion, and a profound respect for the mysteries of the mind.

In the end, the story of OI is not just a tale of scientific discovery but a mirror held up to our own humanity. As we probe the nature of intelligence in a dish, we are forced to confront the fundamental questions of who we are and what makes us truly human.

The journey ahead is sure to be filled with wonder and trepidation in equal measure. But one thing is certain—the era of OI is upon us, and the future will never be the same. As we embark on this adventure, let us do so with open minds, a strong moral compass, and a deep sense of awe at the wonders we are about to uncover. The emergent mind awaits, and the dawn of OI is here.

THANK YOU

Dear Reader,

I would greatly appreciate it if you would please take a moment to share your thoughts on AI Today, OI Tomorrow by leaving a brief review. It takes less than a minute and your verified review not only helps me as a self-publishing author reach more readers, it also helps others like you discover the insights and value the information may offer. Whether it gave you a new perspective or enhanced your understanding of the future of AI and OI, your review is invaluable. I deeply appreciate and thank you for sharing your feedback.

To our future,

Ed

STAY INFORMED

With over thirty years of experience pushing the boundaries of medical technology, Ed is eager to engage with fellow professionals, thought leaders, and innovators. Join him on LinkedIn or visit ai2oi.com to stay updated on his latest insights, projects, and discussions about the future of technology and humanity. Don't miss the opportunity to be part of this exciting journey!

Let's discover the future of intelligence, together.

— Edmund J.B. White | LinkedIn

ABOUT THE AUTHOR

Edmund (Ed) J.B. White is a visionary biomedical innovator, author, and thought leader with over thirty years of experience in pioneering transformative healthcare technologies. With a strong background in biology and medical informatics, he has dedicated his career to pushing the boundaries of what's possible in medical technology.

Ed's journey began when he led the development of groundbreaking electrophysiological technologies for cardiac arrhythmia procedures, saving thousands of lives each year. As the founder and CEO of a biomedical startup, he created the world's first bacterial cellulose artificial graft for coronary bypass surgery, paving the way for lifesaving solutions.

Throughout his career, Ed has consistently been at the forefront of medical innovation, founding successful startups focused on cutting-edge technologies for children with congenital heart disease and digital health solutions for autoimmune disorders. His extensive experience in research, engineering, and commercialization gives him a unique perspective on the emerging field of organoid intelligence (OI).

As an accomplished author, he shares thought-provoking insights into OI, exploring its implications for society and the ethical considerations of technological progress. Through his work, he passionately transforms ideas into innovations that advance healthcare and address the profound questions of humanity's future.

APPENDICIES

APPENDIX A: OI VISIONARIES – SHAPING THE FUTURE

At the heart OI is a transformative vision: to engineer biocomputers from networked 3D cultures of brain cells, or organoids. These aren't just clusters of cells; they're envisioned as the building blocks of systems capable of emulating the human brain's most complex functions—learning, reasoning, and processing sensory information. The fusion of organoids with cutting-edge AI and robotics heralds the creation of hybrid entities that might not only rival but potentially surpass human cognitive abilities. The possibilities are boundless.

Pioneering the Future of OI

This quest for advanced OI is a testament to human ambition and ingenuity. It's a journey that challenges us to innovate across the disciplines of stem cell research, tissue engineering, bioelectronics, and AI. Crafting the optimal OI system architecture raises profound questions:

- Should we channel intelligence through a singular, massive organoid, or harness the collective power of smaller, networked entities?
- How can we best integrate these biological constructs with electronic and robotic systems?
- Is the synthesis of biological and digital networks through advanced algorithms the key to a new form of intelligence?
- Could the secrets to accelerating research lie in the realm of 3D bioprinting of brains?

Solving these puzzles will demand more than isolated effort. It calls for a spirit of open collaboration that transcends borders and disciplines. Only by pooling our collective expertise can we navigate the complexities of this frontier.

OI: A Global Endeavor

Today, the pursuit of OI has quickly become a worldwide endeavor, with leading laboratories and institutions across the globe pushing the frontiers of science and technology. At the forefront of this effort are Professor Thomas Hartung and Professor Lena Smirnova from the Center for Alternatives to Animal Testing at Johns Hopkins University.[112] Their pioneering work focuses on developing innovative, humane testing methods using 3D models of human organs, aiming to understand brain diseases and improve drug development without relying on animal testing. Hartung and Smirnova's approach embodies an ethically responsible path to OI research. Another notable contributor is Professor Alysson Muotri at the University of California, San Diego, whose research on the human social brain is advancing our understanding of brain development and evolution.[113] By utilizing brain organoids derived from stem cells, Muotri's work explores new possibilities for studying neurological disorders, human evolution, and the unique characteristics of the human brain. The drive towards realizing the full potential of OI is marked by a decade of remarkable technological progress. But to unlock the true power of OI, a concerted, global effort is essential. This endeavor requires the formation of elite, multidisciplinary teams that span continents, connecting the dots between academia, industry, and governance.

This collective vision for OI is not just about technological breakthroughs; it's about forging a future where human intelligence and artificial intelligence converge, opening unprecedented possibilities for advancement and understanding.

Visionaries at the Forefront of OI Research

There are several prominent laboratories across the globe pushing the boundaries of science and technology. A selection of these institutions from around the world at the forefront of groundbreaking work include:

The Center for Alternatives to Animal Testing at Johns Hopkins University: Professor Thomas Hartung, Professor Lena Smirnova, and their team are focusing on creating new testing methods that don't use animals. Instead, they're developing tests based on cell behavior and using 3D models of human organs for research, with a particular interest in using brain organoids. They're exploring how these brain models can understand brain diseases better and improve how we develop drugs. Their work is pioneering in moving away from traditional animal testing to more humane and potentially more accurate methods involving OI. Hartung's team is committed to developing OI in an ethically and socially responsible manner by partnering with ethicists from the beginning to establish an "embedded ethics" approach, ensuring all ethical issues are continuously assessed by teams made up of scientists, ethicists, and the public as the research evolves.

Muotri Laboratory at University of California: Professor Alysson R. Muotri, a professor in the Departments of Pediatrics and Cellular & Molecular Medicine at the University of California, San Diego, is working on brain organoids that aim to decipher the complexities of the human social brain, particularly in contexts where social interactions are impaired. By employing human pluripotent stem cells to simulate the neural developmental stages of autism and other social disorders, Muotri creates models of these diseases in a lab setting. This innovative approach enables him to study changes in neuronal networks and test potential drug treatments directly on the organoids. A focus of his research is to identify medications that could restore synaptic connections in conditions like autism, potentially fast-tracking these drugs to clinical trials.

Luxembourg Centre for Systems Biomedicine University of Luxembourg:[114] Professor Jens Schwamborn is specializing in cellular and developmental biology, utilizing organoids as a tool to investigate the onset of neurological disorders such as Parkinson's disease. Schwamborn's team has successfully replicated crucial aspects of the disorder's pathology within an organoid model of Parkinson's, observing both the decline of dopaminergic neurons and the formation of protein aggregates pertinent to the condition. These organoid platforms offer a unique advantage by enabling the examination of Parkinson's disease progression within a network of cells, an approach that traditional monolayer cultures fail to provide.

Emerging Entrepreneurs in OI

There is a wave of excitement permeating through innovative start-ups that could transform the OI landscape. Leading global corporations are also beginning to recognize the immense potential that OI holds.

In early 2023, Amazon's Chief Technology Officer, Werner Vogels, embarked on a journey to visit a pioneering start-up, Cortical Labs, located in Melbourne, Australia.[115] His visit was spurred by a keen interest in exploring ways to reduce data center energy costs. Cortical Labs has garnered significant attention for its groundbreaking work in the realm of biological processing units (BPUs). These BPUs utilize lab-grown neurons to execute computational tasks, a concept that holds tremendous promise. The company's technology made its debut in the scientific journal *Neuron* in October 2022, showcasing that neurons cultivated in a specialized vessels can be coaxed into playing the Atari computer game Pong.[116]

Building upon the success of their DishBrain prototype, Cortical Labs has now developed the CL-1, a more advanced and accessible device that integrates a life-support perfusion circuit with a real-time closed-loop electrophysiological system.[117] The CL-1 represents a significant advancement over its predecessor, the DishBrain, in terms of its ability to interface with biological neural networks. The increased number of stimulation inputs, from eight in the DishBrain to fifty-nine in the CL-1, allows for more complex and diverse patterns of electrical stimulation to be delivered to the cultured neurons. This enhanced stimulation capacity enables researchers to explore a wider range of neural activity patterns and to better understand how these patterns relate to information processing and computation.

Cortical Labs plans to expand the CL-1 to 1,024 stimulation inputs in the coming years, which will provide an unprecedented level of control over the neural network, allowing for the investigation of even more intricate neural dynamics and the potential development of more sophisticated biological computing systems. In addition to the increased number of stimulation inputs, the CL-1 also has a significantly reduced closed-loop latency. Closed-loop latency refers to the time delay between the detection of a neural response and the delivery of a corresponding stimulation input. By achieving a closed-loop latency of less than 1 millisecond (ms), the CL-1 enables real-time interaction between the neural network and the controlling computer system. This near-instantaneous feedback loop is crucial for the development of responsive and adaptive biological computing systems that can learn and process information in real time.

Cortical Labs has also focused on enhancing the stability and accessibility of their technology. The CL-1 is designed to be housed in server racks, allowing well-funded labs to have in-house units or access similar units via the cloud from data center–type facilities. This approach aims to democratize access to synthetic biological intelligence, enabling researchers and developers worldwide to harness the potential of neuromorphic computing. The company plans to have the first server racks operational in their facilities before mid-2025.

The implications of Cortical Labs' work extend far beyond the gaming world. Their innovations have the potential to revolutionize diverse fields, including drug discovery, disease detection, and the development of AI systems capable of continuous learning and adaptation. As the technology continues to evolve, it holds the promise of transforming various industries and shaping the future of computing.

Another company at the forefront of groundbreaking biological computing research is Swiss start-up FinalSpark, founded by neuroscientists Dr. Fred Jordan and Dr. Martin Kutter. FinalSpark is developing a pioneering "wetware" computing platform that utilizes living human brain cells to create machines capable of complex cognition and adaptive behavior.[118] Their approach centers around cultivating brain organoids derived from human stem cells and embedding these brain organoids with tens of thousands of neurons into sophisticated hardware systems.

By connecting the brain organoids to virtual environments through multi-electrode arrays and advanced microfluidics, FinalSpark aims to achieve brain-inspired computing that is vastly more energy-efficient than traditional silicon-based systems. A single organoid containing around 10,000 neurons consumes less than a microwatt of power, orders of magnitude less than even the most efficient AI accelerator chips. But beyond just power savings, FinalSpark believes their brain-inspired systems will ultimately demonstrate higher-order learning, reasoning, and problem-solving abilities that have proven elusive for today's deep learning AI. While still in the early research stages, FinalSpark's wetware computing could pave the way for artificial general intelligence (AGI) that rivals the adaptability and efficiency of the human brain. Their work builds upon research by pioneers like Dr. Steve Potter, who first demonstrated neural control of virtual environments and continues to push the boundaries of what is possible with biological computing substrates.

These enterprising start-ups are pushing the boundaries of what's possible in OI. They not only represent the forefront of technological innovation but also underscore the transformative potential of organoid-interface technologies in reshaping the future of computing and problem-solving.

Concluding Thoughts

At the heart of OI lies a groundbreaking vision: transforming 3D brain cell cultures into biocomputers that mimic the human brain's ability to learn, reason, and process information. This ambitious endeavor seeks to merge organoids with AI and robotics, potentially creating hybrid systems that could outperform human cognitive capabilities.

This audacious quest represents a paradigm shift in our understanding of intelligence and computation. By harnessing the power of organoids, we seek to create machines that not only rival but surpass the capabilities of the human brain. It is a vision that challenges our very notions of what is possible and what the future of technology may hold.

Pioneers in OI are breaking new ground across stem cell research, tissue engineering, bioelectronics, and AI, exploring how to best integrate biological and digital worlds. From global leaders at Johns Hopkins University to innovative start-ups like Cortical Labs and FinalSpark, the race is on to harness the power of organoids in revolutionizing computing, medicine, and AI.

These trailblazers are pushing the boundaries of what we know and what we can achieve. They are exploring uncharted territories, developing new techniques and technologies, and grappling with the profound implications of their work. Their efforts are not just a testament to human ingenuity but also a reflection of our unquenchable thirst for knowledge and progress.

With the backing of initiatives like the US National Science Foundation[119] and a push for international collaboration, OI is becoming a worldwide quest. It is a global effort that transcends borders and disciplines, bringing together the brightest minds from around the world to tackle one of the greatest challenges of our time.

This collaborative spirit is essential to the success of OI. By pooling our resources, sharing our knowledge, and working towards a common goal, we can accelerate the pace of discovery and bring the vision of OI closer to reality. It is a reminder that the greatest achievements of our species have always been the result of collective effort and shared purpose.

ACKNOWLEDGMENTS

This book represents the culmination of a journey that would not have been possible without the dedication, support, and expertise of numerous individuals who contributed their time, insight, and guidance to its creation. As I reflect on the months of research, writing, revising, and collaboration, I am filled with gratitude for the people whose contributions made this work possible. It is with a deep sense of appreciation that I offer these acknowledgments to those who have walked this path with me.

First and foremost, I owe a special debt of gratitude to Erica Ellis at Ink Deep Editing. Erica, your keen editorial instincts and ability to push me to refine my ideas were invaluable. You guided me through the complexities of structuring this book in a way that would resonate with readers, while ensuring clarity and depth throughout. Your tireless efforts to challenge my thinking, combined with your meticulous attention to detail, have left an indelible mark on this manuscript. I deeply appreciate your commitment to excellence and your unwavering support.

I am also deeply thankful to Book Editing Services, who contributed immensely to this book during its early drafts. Your ability to strike a balance between precision and creativity was essential in bringing this project to life. Your team's careful editing ensured that this book maintained the rigor and quality it required at each stage of development.

I would like to express my deepest gratitude to Professors Thomas Hartung and Lena Smirnova from the Center for Alternatives to Animal Testing at Johns Hopkins University. At the forefront of the effort to advance organoid intelligence (OI), your exceptional support and pioneering work have been instrumental in shaping this emerging field. Your collective vision in coining the term OI and building a vibrant community around it has laid the foundation for future explorations. The establishment of the organoid intelligence website (https://organoidintelligence.org) and newsletter under your leadership has provided an invaluable resource for researchers, technologists, ethicists, and policymakers. Your dedication to advancing OI has been both inspiring and transformative, and I am honored to have benefited from your insights.

This book greatly benefited from the wisdom and expertise of several subject matter experts who contributed to my understanding of the ethical, spiritual, and scientific dimensions explored in this work. Daniel Burger, your innovative thinking and creative input provided a framework for many of the core ideas that helped shape the direction of this book. Your perspectives on the future of technology were inspiring during critical moments, allowing me to push the boundaries of my own thinking. I would also like to acknowledge Dr. Ewelina Kurtys, whose contributions enriched the narrative. I am grateful for your support.

I am deeply appreciative of Professor Andrea Lavazza, who reviewed my text and offered thoughtful feedback. Your philosophical insights, particularly regarding the interpretation of organoid behavior and consciousness, prompted me to refine my arguments. Your suggestion to reconsider how we discuss organoid behavior enhanced the clarity and accuracy of this work. Thank you for your valuable input, which helped strengthen the ethical discussions presented in this book.

I would also like to extend my sincere appreciation to Dr. Brett Kagan for his invaluable input to this book. His team's groundbreaking work in creating biological neural networks and exploring the intersection of neuroscience and artificial intelligence provided crucial insights into the future of brain–computer interfaces. His innovative approach and cutting-edge research significantly enhanced the depth and relevance of the technological discussions in this work.

Very Reverend Albert Bogle, your profound reflections on the intersection of faith and emerging technologies added a rich theological dimension to this book. Our discussions on organoid intelligence and its alignment with the Christian values of love and grace provided a deep moral lens through which I could explore these ethical quandaries. Your thought-provoking words, particularly around the tension between the promise of technological advancements and the potential perils of human choices, helped ground this work in meaningful reflection.

Honorable Rabbi Daniel Nevins, your contributions on the Jewish perspective of brain organoid research were instrumental in broadening my understanding of the ethical landscape. Your insights on the sanctity of life and humanity's role as partners in the divine act of healing illuminated the ways in which Jewish tradition may engage with the complexities of this research. Your reflections, especially around the concept of personhood and how it applies to scientific advances, provided critical depth to the discussions presented in this book.

To Dr. Pete Philips, I extend my thanks for connecting me with scholars who stand at the fascinating nexus of technology, biology, spirituality, and ethics. Your introductions opened doors to perspectives I would not have otherwise encountered, enriching the discourse and expanding the scope of this work.

A special thank you to Danielle Rosales from Space Tango for granting me permission to share images from earlier space missions, which provided a visual element that brought key ideas to life for readers. Your support in allowing me to incorporate these images helped illustrate complex concepts that words alone could not fully capture, and I appreciate your collaboration in this effort.

Dr. Shai Tubali, your deep insights into spirituality, consciousness, and personal transformation provided a vital layer to the narrative. Your contributions helped me explore the ethical implications of technology from a more introspective and philosophical perspective. Your work enriched this book in ways that I hope resonate with readers both intellectually and emotionally.

I also want to thank Professor Alysson Muotri for generously sharing your groundbreaking research and images from your work in stem cell orbital research. Your pioneering studies in this field brought new dimensions to my exploration of human potential in space environments, grounding theoretical ideas in tangible science. The intellectual depth and visual clarity you provided helped make complex scientific concepts accessible to readers.

A special mention goes to Dr. Chris Smith, whose expertise on open innovation in pharmaceutical research provided a critical lens for the scientific discussions in this book. I am deeply grateful for your contributions.

Lisa Park, your innovative installations that translate brainwave activity into immersive audiovisual experiences provided a fascinating exploration of the mind's creative capacity. Though your work doesn't directly involve organoids, it draws inspiration from cutting-edge research in this field and serves as a powerful metaphor for the potential of organoid intelligence. I am grateful for your contribution, which helped deepen the discussion on the intersection of art and technology in this book.

I am particularly grateful to Katja Rausch who has been a wonderful sounding board for ideas. Katja's extensive background in new technologies and ethics, combined with the House of Ethics' diverse global perspectives, provided unique insights that significantly enriched the ethical discussions in this book.

I am also incredibly fortunate to have had the support of talented proofreaders. Sercan Deniz, Dr. Lorena R. Lizarraga Valderrama, John Bibby, Oliver Stallwood and Prof. Dr. Bonaventure Soh Bejeng Ndikung, thank you for your careful attention to detail and for catching the nuances that make all the difference. Your dedication to ensuring accuracy and clarity helped shape this book into its final form, and I deeply appreciate your hard work.

I would like to express my sincere gratitude to Mike Trent, the brilliant designer behind the book cover. Your artistic vision and ability to encapsulate the essence of this book into a captivating cover have greatly enhanced its appeal. Your creativity and attention to detail have provided a visual gateway that invites readers into the world within these pages. Thank you for your exceptional contribution.

To Archangel Ink Publishers, I extend my sincere thanks for your proactive approach throughout the publishing process. Your constant support, guidance, and professionalism made the journey from draft to publication as seamless as possible. Your team's commitment to excellence ensured that this book would reach its full potential, and I am thankful for the care you took in bringing this project to completion.

On a personal level, I owe a debt of gratitude to my family. Thank you for your unwavering support, patience, and encouragement throughout this long process. You endured endless conversations about the book, offered valuable advice, and gave me the time and space I needed to write. Your belief in me was a constant source of motivation during the most challenging moments, and I cannot thank you enough for your love and support.

To my friends, thank you for being my sounding boards, offering honest feedback, and giving me much-needed breaks from the intense writing process. Your willingness to listen, critique, and encourage me at every stage of this journey was invaluable. Your insights often proved as valuable as those of the experts I consulted, and your emotional support kept me going.

This book is a reflection of the collective wisdom, insights, and contributions of these incredible individuals as much as it is my own. Without your guidance, feedback, and support, this book would not have become what it is today. From the bottom of my heart, thank you for helping me bring this vision to life

LIST OF ILLUSTRATIONS

The author conveys appreciation for the permission to use the images featured in this book. Every effort has been made to communicate with all copyright owners, and any discrepancies brought to the author's notice will be rectified in forthcoming editions.

"Eunoia"
Credit: Courtesy of the Artist
© Lisa Park

Professor Christof Koch
Credit: Erik Dinnell, Allen Institute
© Allen Institute

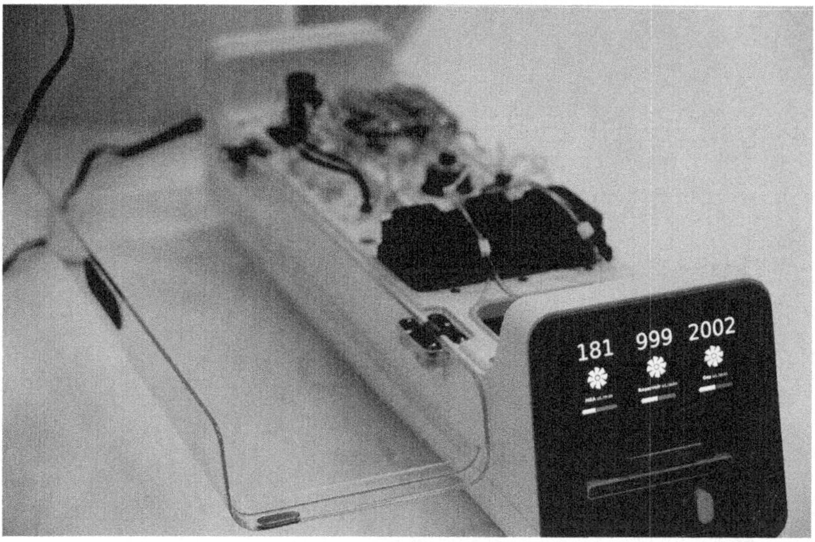

CL1 Unit - Integrates life-support perfusion circuit with a real-time closed-loop electrophysiological system
Credit: Cortical Labs Pte Ltd
© Cortical Labs Pte Ltd

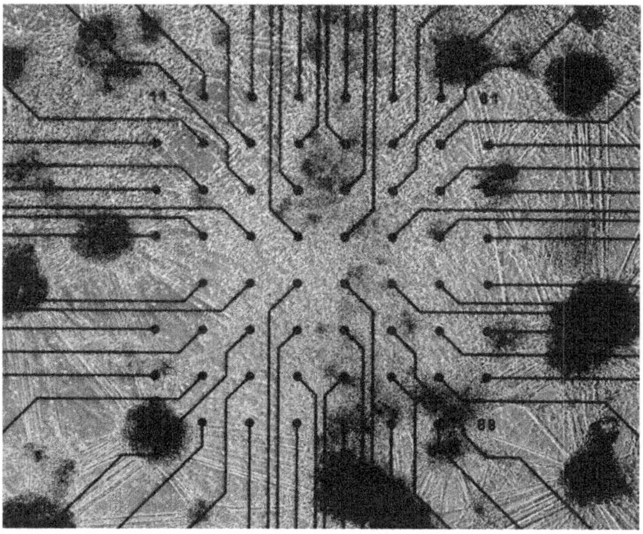

Cells with Clusters
Credit: Cortical Labs Pte Ltd
© Cortical Labs Pte Ltd

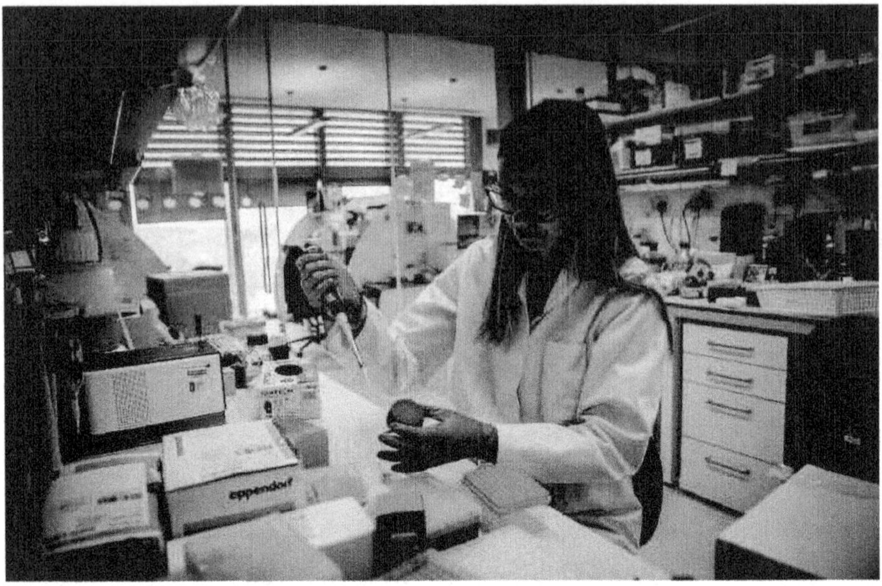

Professor Madeline Lancaster
Credit: Lloyd Mann and the University of Cambridge

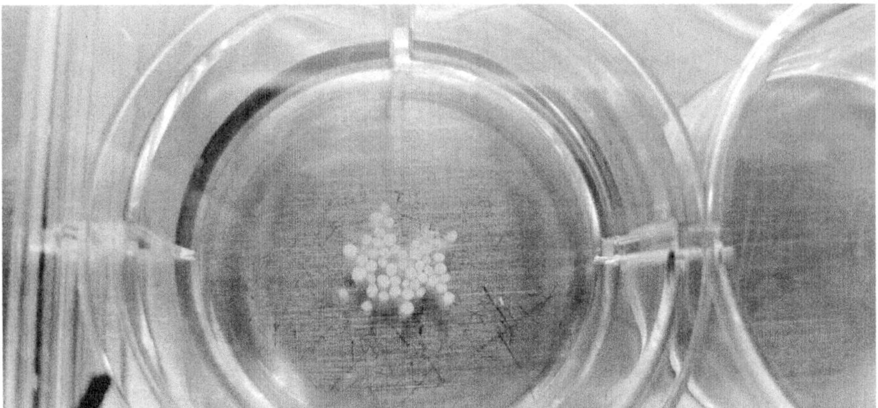

Human Brain Organoids
Credit: Muotri Lab, UC San Diego

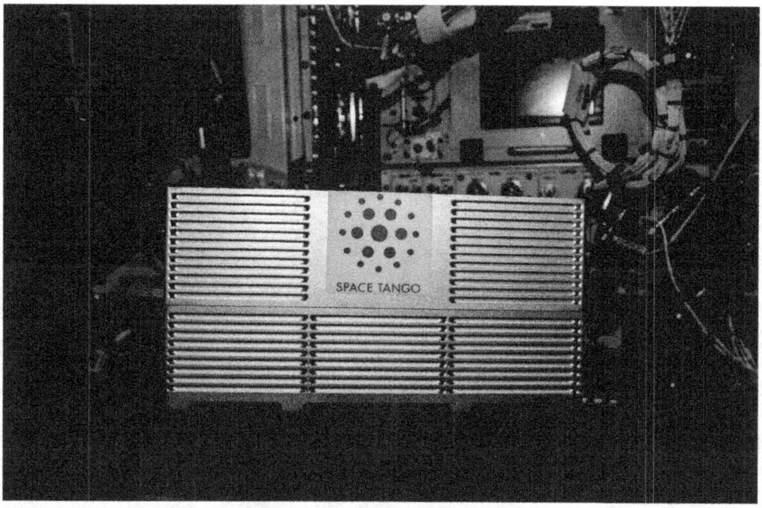

CubeLabs: Brain organoids housed inside autonomous hardware on the International Space Station
Credit: Courtesy of Space Tango
© Space Tango

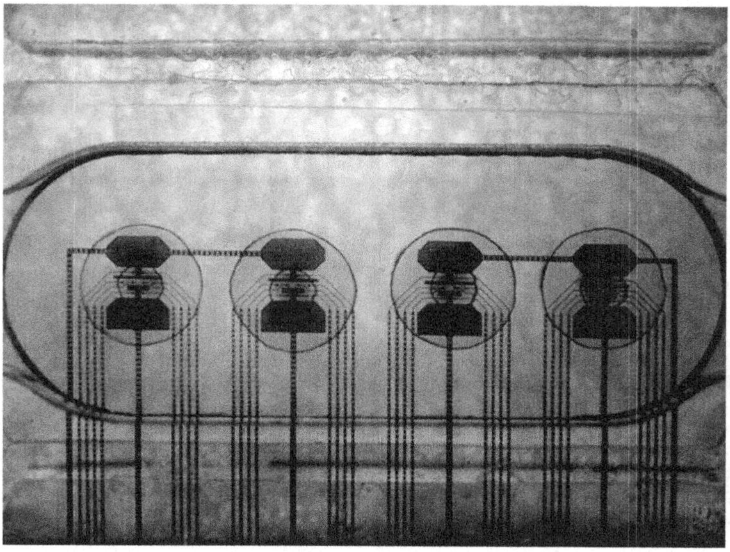

Four human brain organoids, each with around 10,000 living human brain cells, wired into a biocomputing array in FinalSpark's Neuroplatform
Credit: FinalSpark
© FinalSpark

Professor Alysson Muotri
Credit: Muotri Lab, UC San Diego

Professor Thomas Hartung and Assistant Professor Lena Smirnova
with a picture of an organoid
Credit: Will Kirk, Johns Hopkins University

Dr. Fred Jordan and Dr. Martin Kutter
Credit: FinalSpark
© FinalSpark

Sculpture "Seven Generations" by Frederick Franck
Credit: Photo by Mabit1
Licensed under CC BY-SA 3.0
License URL: https://creativecommons.org/licenses/by-sa/3.0/

Representatives at the Biological Weapons Convention
Credit: Eric Bridiers, US Mission Geneva

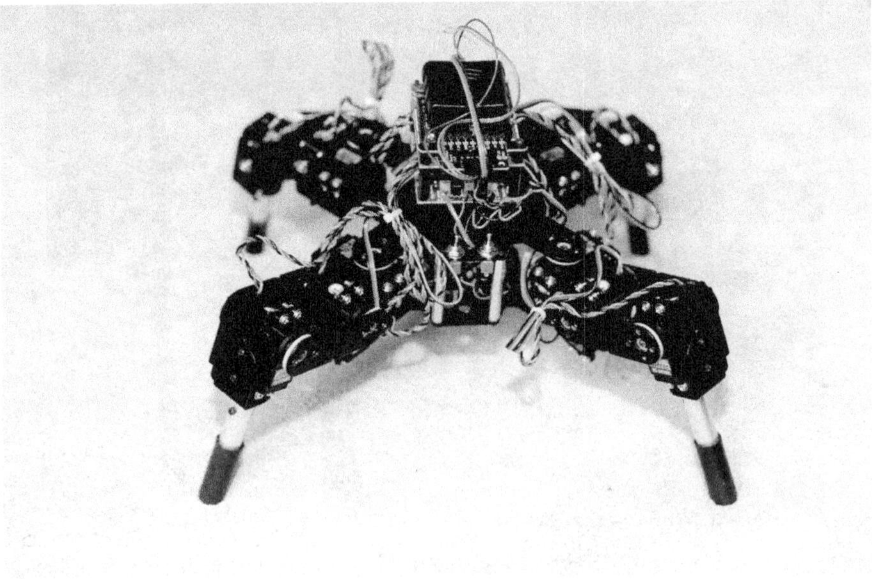

Cyborg Insect
Credit: Muotri Lab, UC San Diego

GLOSSARY OF KEY TERMS

3D Bioprinting: An innovative technology used to create three-dimensional structures from living cells, with the potential to create more complex and functional brain organoids.

3D Culture: A method of growing cells in a three-dimensional environment, which is essential for the formation of organoids.

Advanced Microscopy: Techniques that provide high-resolution images of biological samples, essential for studying the detailed structures within brain organoids.

Artificial General Intelligence (AGI): A level of artificial intelligence where a machine can understand, learn, and apply intellectual abilities across a wide range of tasks, similar to human intelligence.

Artificial Neural Networks (ANN): Computer systems inspired by the human brain, consisting of interconnected nodes that process information and learn from data to perform tasks.

Assembloids: 3D cell cultures that combine different lab-grown mini-organs or tissues, allowing researchers to study complex biological processes, interactions between tissues, and human organ systems more accurately than simpler models.

Astrocyte: A star-shaped glial cell in the brain and spinal cord that supports and nourishes neurons, helps maintain brain health, and plays a role in communication between brain cells.

Attention Deficit Hyperactivity Disorder (ADHD): A neurodevelopmental disorder characterized by problems with paying attention, excessive activity, or difficulty controlling behavior.

Biocomputers: Systems that integrate biological components, such as brain cells or organoids, with electronic computing elements to perform computational tasks.

Bioelectronics: A scientific field that combines biology and electronics to create devices that interact with biological systems, crucial for integrating organoids with electronic and robotic systems.

Bioethics: The study of ethical issues arising from advances in biology and medicine.

Biohybrid Interfaces: Technological interfaces that integrate biological and artificial components, potentially allowing for direct interaction between brain organoids and machines.

Biological Neural Networks (BNN): Interconnected systems of neurons in the brain and nervous system that process, transmit, and receive information to enable learning, perception, and behavior.

Biological Processing Units (BPUs): Innovative computational devices that utilize living neurons to perform calculations, offering a new approach to processing information with high efficiency.

Bioreactors: Devices that support a biologically active environment, used in organoid cultivation to mimic the body's conditions.

Boolean Algebra: A branch of mathematics that operates on binary variables and is fundamental to the design of digital computer circuits.

Brain Organoids: A specific type of organoid designed to replicate aspects of brain structure and function.

Brainoware: A conceptual fusion of biological brain matter with hardware, aiming to create computing systems that mimic the brain's functionality and efficiency.

Closed-loop latency: The time delay between the moment a system detects an event or receives an input and the moment it generates a corresponding response or output.

Cognition: The mental action or process of acquiring knowledge and understanding through thought, experience, and the senses.

Cosmic Conservation: The concept of preserving the natural state of outer space environments, similar to conservation efforts on Earth.

Cyborg Insects: Insects integrated with electronic controls that allow their movements to be manipulated; used as models for testing neural control mechanisms.

Dementia: A broad category of brain diseases that cause a long-term and often gradual reduction in cognitive abilities, affecting daily functioning.

Differentiation: The process by which a cell changes from one type to another, typically becoming a more specialized type.

Dual-Use Dilemma: The issue that arises when technologies can be used for both beneficial purposes and harmful ones.

Electrophysiological Signaling: The electrical activity of cells, particularly neurons, which is crucial for brain function.

Embedded Ethics: An approach that integrates ethical considerations directly into the research and development process of new technologies, ensuring that ethical issues are addressed as part of the innovation cycle.

Embryonic Stem Cells: Cells derived from the early stages of embryo development that have the potential to develop into any cell type in the body.

Entanglement: A quantum phenomenon where particles become interconnected so that the state of one can instantly affect the state of another, regardless of the distance between them.

Ethical Considerations: The evaluation of moral principles and values that govern the conduct of activities, especially regarding their impact on sentient beings.

Gene Expression: The process by which the instructions in DNA are converted into a functional product, such as a protein; fundamental in determining the type and role assumed by a cell during differentiation.

Glial Cell: A type of brain cell that supports, protects, and nourishes neurons, helping to maintain brain health and function.

Global Workspace Theory (GWT): A cognitive architecture theory suggesting that consciousness arises from the broadcasting of information to a "global workspace" in the brain, making it available for various processes.

Growth Factors: Natural substances capable of stimulating cellular growth, proliferation, healing, and differentiation.

In Vitro: Latin for "in glass," referring to biological processes carried out outside a living organism, typically in a lab environment.

Induced Pluripotent Stem Cells (iPSCs): A type of stem cell generated directly from adult cells that can differentiate into various types of cells.

Integrated Information Theory (IIT): A theoretical framework for understanding consciousness; posits that consciousness corresponds to the capacity of a system to integrate information.

Microfluidics: Technology that processes or manipulates small amounts of fluids, used to control the environment around cells and organoids.

Microgravity: The condition in which people or objects appear to be weightless, as experienced in orbiting spacecraft.

Morphogenesis: The biological process that causes an organism to develop its shape; critical in the formation of organoids' structure.

Microsphysiology Systems (MPS): Advanced laboratory models that mimic the structure and function of human brain tissue on a miniature scale, using three-dimensional cell cultures called organoids.

Nanoscale Biosensors: Tiny sensors capable of detecting and measuring biological phenomena at the nanometer scale, crucial for interfacing with brain organoids.

Nanowire Arrays: Semiconductor devices used in sensors and electronics, where nanowires are used as sensitive elements for detecting signals from biological tissues.

Neural Networks: Complex networks of neurons responsible for transmitting and processing information in the brain.

Neural Tube: The embryonic precursor to the central nervous system, which includes the brain and spinal cord.

Neurodiversity: Recognition of the diversity of human brains and minds, stressing the importance of including various cognitive abilities and disorders in research and applications.

Neuroethics: The field of study concerned with the ethical, legal, and social implications of neuroscience.

Neurogenesis: The process by which neurons are generated from neural stem cells and progenitor cells.

Neuromodulation: The process of regulating neural activity, often using chemicals or electrical stimulation, to influence the behavior of neural circuits.

Neuron Doctrine: The concept that the nervous system is made up of discrete individual cells, a principle that has implications for understanding how the brain processes information.

Non-Maleficence: The principle of doing no harm, especially relevant in the context of research ethics.

Oligodendrocyte: A type of glial cell in the brain and spinal cord that wraps around neurons, forming an insulating layer called myelin, which helps neurons transmit signals quickly and efficiently.

Optogenetics: A technique in neuroscience where genes for light-sensitive proteins are introduced into specific types of neurons to control their activity using light.

Organoid Intelligence (OI): An emerging multidisciplinary field aiming to develop biological computing using 3D cultures of human brain cells, also known as brain organoids. The vision of organoid intelligence is to engineer biocomputers from networked brain organoids that can mimic complex brain functions, such as learning, reasoning, and processing sensory information.

Organoid Neural Networks (ONNs): Networks formed by connecting brain organoids to electronic interfaces, allowing them to process information in a manner analogous to biological brains but with significantly reduced energy consumption.

Organoids: Lab-grown 3D structures made from stem cells that mimic the complexity of an organ.

Personhood and Moral Status: The debate over whether entities like brain organoids possess qualities that would warrant ethical consideration equivalent to that extended to living beings.

Phenotype: The set of observable characteristics or traits of an organism, such as morphology, development, or behavior, resulting from the interaction of its genotype with the environment.

Progenitor Cells: Cells that tend to differentiate into a specific type of cell but are more limited than stem cells in the types of cells they can become.

Quantum Computing: A type of computing that utilizes quantum-mechanical phenomena, such as superposition and entanglement, to perform operations on data.

Radical Enhancement: Significant improvements or modifications to biological entities or systems, potentially exceeding natural capabilities.

Sentience: The capacity to feel, perceive, or experience subjectively.

Singularity: A hypothetical future point at which technological growth becomes uncontrollable and irreversible, resulting in unforeseeable changes to human civilization, often associated with the advent of AGI surpassing human intelligence.

Space Ethics: The study of moral, ethical, and legal issues surrounding human activities in outer space.

Space Medicine: The practice of medicine that studies the health of astronauts living in outer space.

Specialized Vessel: A culture system specifically designed to support the growth and development of three-dimensional brain tissue structure.

Stem Cells: Cells with the potential to develop into many different cell types in the body during early life and growth.

Superposition: The ability of a quantum system, like a qubit, to exist in multiple states at once. This allows quantum computers to perform many calculations simultaneously.

Sustainable AI: Artificial intelligence research and applications designed to minimize energy consumption and reduce environmental impact, aligning with broader goals of sustainability.

Synaptic Connections: The junctions between neurons that allow them to communicate with each other.

Ultrasound: Acoustic waves with frequencies higher than the upper limit of human hearing, used in medical imaging and treatments, and explored for stimulating specific regions in the brain.

Virtual Playpens: Controlled digital environments designed to limit the influence or reach of an AI or organoid intelligence.

ENDNOTES

Introduction

1 Kagan, B. J., Kitchen, A. C., Tran, N. T., Rollo, B., Razi, A., & Friston, K. J. (2022). The DishBrain system: A real-time synthetic biological intelligence platform. *Neuron, 110*(24), 3952–3969. https://doi.org/10.1016/j.neuron.2022.09.001

2 Smirnova, L., Caffo, B. S., Gracias, D. H., Huang, Q., Morales Pantoja, I. E., Tang, B., Zack, D. J., Berlinicke, C. A., Boyd, J. L., Harris, T. D., Johnson, E. C., Kagan, B. J., Kahn, J., Muotri, A. R., Paulhamus, B. L., Schwamborn, J. C., Plotkin, J., Szalay, A. S., Vogelstein, J. T., … Hartung, T. (2023). Organoid intelligence (OI): The new frontier in biocomputing and intelligence-in-a-dish. *Frontiers in Science*, *1*, Article 1017235. https://doi.org/10.3389/fsci.2023.1017235

3 Paşca, S. P., Arlotta, P., Bateup, H. S., Camp, J. G., Cappello, S., Gage, F. H., Knoblich, J. A., Kriegstein, A. R., Lancaster, M. A., Ming, G.-L., Muotri, A. R., Park, I.-H., Reiner, O., Song, H., Studer, L., Temple, S., Testa, G., Treutlein, B., & Vaccarino, F. M. (2022). A nomenclature consensus for nervous system organoids and assembloids. *Nature*, *609*(7929), 907–910. https://doi.org/10.1038/s41586-022-05219-6

Chapter 1: Unveiling Brain Organoid Basics

4 Lancaster, M. A., & Knoblich, J. A. (2014). Organogenesis in a dish: Modeling development and disease using organoid technologies. *Science, 345*(6194), 1247125. https://doi.org/10.1126/science.1247125

5 Takahashi, K., & Yamanaka, S. (2006). Induction of pluripotent stem cells from mouse embryonic and adult fibroblast cultures by defined factors. *Cell*, *126*(4), 663–676. https://doi.org/10.1016/j.cell.2006.07.024

6 Chambers, S. M., Fasano, C. A., Papapetrou, E. P., Tomishima, M., Sadelain, M., & Studer, L. (2009). Highly efficient neural conversion of human ES and iPS cells by dual inhibition of SMAD signaling. *Nature Biotechnology*, *27*(3), 275–280. https://doi.org/10.1038/nbt.1529

7 Sasai, Y. (2013). Next-generation regenerative medicine: organogenesis from stem cells in 3D culture. *Cell Stem Cell*, *12*(5), 520-530. https://doi.org/10.1016/j.stem.2013.04.009

8 Qian, X., Nguyen, H.N., Song, M.M., et al. (2016). Brain-region-specific organoids using mini-bioreactors for modeling ZIKV exposure. *Cell*, *165*(5), 1238–1254. https://doi.org/10.1016/j.cell.2016.04.032

9 Bhatia, S.N., Ingber, D.E. (2014). Microfluidic organs-on-chips. *Nat Biotechnol*, *32*(8), 760–772. https://doi.org/10.1038/nbt.2989

10 Quadrato, G., et al. (2017). Cell diversity and network dynamics in photosensitive human brain organoids. *Nature*, *545*(7652), 48–53. https://doi.org/10.1038/nature22047

11 Firth, A.L., Menon, T., Parker. G.S., et al. (2015). Functional gene correction for cystic fibrosis in lung epithelial cells generated from patient iPSCs. *Cell Rep*, *12*(9), 1385–1390. https://doi.org/10.1016/j.celrep.2015.07.062

Chapter 2: Realizing OI Potential

12 Mim, M. S., Knight, C., & Zartman, J. J. (2023). Quantitative insights in tissue growth and morphogenesis with optogenetics. *Physical Biology*, *20*(6), 061001. https://doi.org/10.1088/1478-3975/acf7a1

13 Løvmo, M. K., Deng, S., Moser, S., Leitgeb, R., Drexler, W., & Ritsch-Marte, M. (2024). Ultrasound-induced reorientation for multi-angle optical coherence tomography. *Nature Communications*, *15*, Article 2391. https://doi.org/10.1038/s41467-024-46506-2

14 Park, S. B., Koh, B., Kwon, H. S., Kim, Y. E., Kim, S. S., Cho, S.-H., Kim, T.-Y., Bae, M. A., Kang, D., Kim, C. H., & Kim, K. Y. (2023). Quantitative and qualitative analysis of neurotransmitter and neurosteroid production in cerebral organoids during differentiation. *ACS Chemical Neuroscience*, *14*(20), 3761–3771. https://doi.org/10.1021/acschemneuro.3c00246

15 Qian, X., Song, H., & Ming, G. (2019). Brain organoids: Advances, applications and challenges. *Development*, *146*(8), dev166074. https://doi.org/10.1242/dev.166074

16 Garita-Hernandez, M., Guibbal, L., Toualbi, L., Routet, F., Chaffiol, A., Winckler, C., Harinquet, M., Robert, C., Fouquet, S., Bellow, S., Sahel, J. A., Goureau, O., Duebel, J., & Dalkara, D. (2018). Optogenetic light sensors in human retinal organoids. *Frontiers in Neuroscience*, *12*, 789. https://doi.org/10.3389/fnins.2018.00789

Chapter 4: Synapse and Silicon: Decoding Neural Networks

17 Castiglione, H., Vigneron, P.-A., Baquerre, C., Yates, F., Rontard, J., & Honegger, T. (2022). Human brain organoids-on-chip: Advances, challenges, and perspectives for preclinical applications. *Pharmaceutics*, *14*(11), 2301. https://doi.org/10.3390/pharmaceutics14112301

18 Saglam-Metiner, P., Yildirim, E., Dincer, C., Gupta, A., Ayan, B., Serhatli, E., Keles, H. O., Erkoc, P., Yildiz, I., Sencan, I., Isik, S., Tasoglu, S., & Guler, M. O. (2024). Humanized brain organoids-on-chip integrated with sensors for screening neuronal activity and neurotoxicity. *Microchimica Acta*, *191*(1), 71. https://doi.org/10.1007/s00604-023-06165-4

19 Kagan, B. J., Kitchen, A. C., Tran, N. T., Rollo, B., Razi, A., & Friston, K. J. (2022). The DishBrain system: A real-time synthetic biological intelligence platform. *Neuron, 110*(24), 3952–3969. https://doi.org/10.1016/j.neuron.2022.09.001

20 Angello, N. H., Friday, D. M., Hwang, C., Yi, S., Cheng, A. H., Torres-Flores, T. C., Jira, E. R., Wang, W., Aspuru-Guzik, A., Burke, M. D., Schroeder, C. M., Diao, Y., & Jackson, N. E. (2024). Closed-loop transfer enables artificial intelligence to yield chemical knowledge. *Nature, 633*(7979), 351–358. https://doi.org/10.1038/s41586-024-07892-1

Chapter 5: Insights from Lab-Grown Minds

21 Scott, G., & Huang, Y. (2022). Engineering cerebral folding in brain organoids. *Neural Regeneration Research, 17*(11), 2420–2422. https://doi.org/10.4103/1673-5374.335789

22 Burrows, L. (2016, February 1). How, not why, the human brain folds. *The Harvard Gazette, SEAS Communications*. https://news.harvard.edu/gazette/story/2016/02/how-not-why-the-human-brain-folds/

Chapter 6: Medical Revolution: Dish-Derived Healing

23 DiMasi, J. A., Grabowski, H. G., & Hansen, R. W. (2016). Innovation in the pharmaceutical industry: New estimates of R&D costs. *Journal of Health Economics, 47*, 20–33. https://doi.org/10.1016/j.jhealeco.2016.01.012

24 Zhou, J.-Q., Zeng, L.-H., Li, C.-T., He, D.-H., Zhao, H.-D., Xu, Y.-N., Jin, Z.-T., & Gao, C. (2023). Brain organoids are new tool for drug screening of neurological diseases. *Neural Regeneration Research, 18*(9), 1884–1889. https://doi.org/10.4103/1673-5374.367983

25 Jensen, G., Morrill, C., & Huang, Y. (2018). 3D tissue engineering, an emerging technique for pharmaceutical research. *Acta Pharmaceutica Sinica B, 8*(5), 756–766. https://doi.org/10.1016/j.apsb.2018.03.006

26 Karmirian, K., Holubiec, M., Goto-Silva, L., Fernandez Bessone, I., Vitória, G., Mello, B., Alloatti, M., Vanderborght, B., Falzone, T. L., & Rehen, S. (2023). Modeling Alzheimer's disease using human brain organoids. In Chun, J. (ed*), Methods in Molecular Biology* (Vol. 2561, pp. 135–158). Springer. https://doi.org/10.1007/978-1-0716-2655-9_7

27 Lau, J. (2022, February 2). Different autism risk genes, same effects on brain development. *Harvard Stem Cell Institute*. https://hsci.harvard.edu/news/autism-risk-genes-brain-development-organoids

Chapter 7: Neurological Clarity: Role of Lab Neurons

28 Raja, W. K., Mungenast, A. E., Lin, Y. T., Ko, T., Abdurrob, F., Seo, J., & Tsai, L. H. (2016). Self-organizing 3D human neural tissue derived from induced pluripotent stem cells recapitulate Alzheimer's disease phenotypes. *PLoS ONE*, *11*(9), e0161969. https://doi.org/10.1371/journal.pone.0161969

29 Lin, Y. T., Seo, J., Gao, F., Feldman, H. M., Wen, H. L., Penney, J., Cam, H. P., Gjoneska, E., Raja, W. K., Cheng, J., Rueda, R., Kritskiy, O., Abdurrob, F., Peng, Z., Milo, B., Yu, C. J., Elmsaouri, S., Dey, D., Ko, T., Yankner, B. A., & Tsai, L. H. (2018). APOE4 causes widespread molecular and cellular alterations associated with Alzheimer's disease phenotypes in human iPSC-derived brain cell types. *Neuron*, *98*(6), 1141–1154.e7. https://doi.org/10.1016/j.neuron.2018.05.008

30 Ghatak, S., Dolatabadi, N., Trudler, D., Zhang, X., Wu, Y., Mohata, M., Ambasudhan, R., Talantova, M., & Lipton, S. A. (2019). Mechanisms of hyperexcitability in Alzheimer's disease hiPSC-derived neurons and cerebral organoids vs isogenic controls. *eLife*, *8*, e50333. https://doi.org/10.7554/eLife.50333

31 Zhou, J.-Q., Zeng, L.-H., Li, C.-T., He, D.-H., Zhao, H.-D., Xu, Y.-N., Jin, Z.-T., & Gao, C. (2023). Brain organoids are new tool for drug screening of neurological diseases. *Neural Regeneration Research*, *18*(9), 1884–1889. https://doi.org/10.4103/1673-5374.367983

32 Skaria, A. P. (2022). The economic and societal burden of Alzheimer disease: managed care considerations. *Am J Manag Care*. 2022 Sep;28(10 Suppl): S188-S196. https://doi.org/10.37765/ajmc.2022.89236

33 Stachowiak, E. K., Benson, C. A., Narla, S. T., Dimitri, A., Bayona Chuye, L. E., Dhiman, S., Harikrishnan, K., Elahi, S., Freedman, D., Brennand, K. J., Sarder, P., & Stachowiak, M. K. (2017). Cerebral organoids reveal early cortical maldevelopment in schizophrenia—computational anatomy and genomics, role of FGFR1. *Translational Psychiatry*, 7, Article 6. https://www.nature.com/articles/s41398-017-0054-x

34 Kathuria, A., Lopez-Lengowski, K., Jagtap, S. S., McPhie, D., Perlis, R. H., Cohen, B. M., & Karmacharya, R. (2020). Transcriptomic landscape and functional characterization of induced pluripotent stem cell-derived cerebral organoids in schizophrenia. *JAMA Psychiatry*, *77*(7), 745–754. https://doi.org/10.1001/jamapsychiatry.2020.0196

35 Stachowiak, E.K., Benson, C.A., Narla, S.T., et al. (2017). Cerebral organoids reveal early cortical maldevelopment in schizophrenia-computational anatomy and genomics, role of FGFR1. *Transl Psychiatry*, 7(11), 6. https://doi.org/10.1038/s41398-017-0054-x

36 Notaras, M., Lodhi, A., Dündar, F., Collier, P., Sayles, N. M., Tilgner, H., Greally, J. M., Gage, F. H., & Kushner, S. A. (2022). Schizophrenia is defined by cell-specific neuropathology and multiple neurodevelopmental mechanisms in patient-derived cerebral organoids. *Molecular Psychiatry*, *27*(3), 1416–1434. https://doi.org/10.1038/s41380-021-01316-6

37 Nascimento, J.M., Saia-Cereda, V.M., Zuccoli, G.S., et al. (2022). Proteomic signatures of schizophrenia-sourced iPSC-derived neural cells and brain organoids are similar to patients' postmortem brains. *Cell Biosci.*, *12*, 189. https://doi.org/10.1186/s13578-022-00928-x

38 Villanueva, R. (2023). Advances in the knowledge and therapeutics of schizophrenia, major depression disorder, and bipolar disorder from human brain organoid research. *Frontiers in Psychiatry*, *14*, Article 1178494. https://doi.org/10.3389/fpsyt.2023.1178494

39 Zhang, K., Yu, F., Zhu, J., Han, S., Chen, J., Wu, X., Chen, Y., Shen, T., Liao, J., Guo, W., Yang, X., Wang, R., Qian, Y., Yang, J., Cheng, L., Zhao, Y., Hui, C.-C., Li, J., Peng, G., He, S., & Tang, K. (2020). Imbalance of excitatory/inhibitory neuron differentiation in neurodevelopmental disorders with an NR2F1 point mutation. *Cell Reports*, *31*(3), 107521. https://doi.org/10.1016/j.celrep.2020.03.085

40 Li, C., Fleck, J. S., Martins-Costa, C., Burkard, T. R., Themann, J., Stuempflen, M., Peer, A. M., Vertesy, Á., Littleboy, J. B., Esk, C., Elling, U., Kasprian, G., Corsini, N. S., Treutlein, B., & Knoblich, J. A. (2023). Single-cell brain organoid screening identifies developmental defects in autism. *Nature*, *621*, 373–380. https://doi.org/10.1038/s41586-023-06473-y

41 Chan, W. K., Griffiths, R., Price, D. J., & Mason, J. O. (2020). Cerebral organoids as tools to identify the developmental roots of autism. *Molecular Autism*, *11*, Article 58. https://doi.org/10.1186/s13229-020-00360-3

42 Kim, S., & Chang, M. (2023). Application of human brain organoids—Opportunities and challenges in modeling human brain development and neurodevelopmental diseases. *International Journal of Molecular Sciences*, *24*(15), 12528. https://doi.org/10.3390/ijms241512528

43 Goldrick, C., Guri, I., Herrera-Oropeza, G., O'Brien-Gore, C., Roy, E., Wojtynska, M., Cóndor, M., Gispert-Sánchez, S., & Metzakopian, E. (2023). 3D multicellular systems in disease modelling: From organoids to organ-on-chip. *Frontiers in Cell and Developmental Biology*, *11*, Article 1083175. https://doi.org/10.3389/fcell.2023.1083175

44 Dinan, T. G., & Cryan, J. F. (2012). Regulation of the stress response by the gut microbiota: Implications for psychoneuroendocrinology. *Psychoneuroendocrinology*, *37*(9), 1369–1378. https://doi.org/10.1016/j.psyneuen.2012.03.007

45 Goswami, A., Wendt, F.R., Pathak, G.A., Volkmar, F.R., Lipska, B.K., Polimanti, R. (2021). Role of microbes in the pathogenesis of neuropsychiatric disorders. *Prog Mol Biol Transl Sci.*, *176*, 133–153. https://doi.org/10.1016/j.yfrne.2021.100917

46 Passaro, A. P., & Stice, S. L. (2021). Electrophysiological analysis of brain organoids: Current approaches and advancements. *Frontiers in Neuroscience, 14*, 622137. https://doi.org/10.3389/fnins.2020.622137

Chapter 8: Memory in Brain Organoids: Fact or Fiction?

47 Osaki, T., & Ikeuchi, Y. (2021). Complex activity and short-term memories in reciprocally connected cerebral organoids. bioRxiv. https://doi.org/10.1101/2021.02.16.431387

48 Suddendorf, T., Addis, D. R., & Corballis, M. C. (2009). Mental time travel and the shaping of the human mind. *Philosophical Transactions of the Royal Society B: Biological Sciences, 364*(1521), 1317–1324. https://doi.org/10.1098/rstb.2008.0301

Chapter 9: Brain Organoids: Unveiling Sleep and Dreams

49 Born, J., & Wilhelm, I. (2012). System consolidation of memory during sleep. *Psychological Research, 76*(2), 192–203. https://doi.org/10.1007/s00426-011-0335-6

50 American Gene & Cell Therapy Society. (2023, July 5). Psychedelics are making a therapeutic comeback, and new research is demystifying how they work. *AGS Therapeutics Publications*. https://www.ags-tx.com/publications/psychedelics-therapeutic-comeback-new-research-demystifying-how-they-work

51 Neuroskeptic. (2019, September 7). Brainwaves in organoids? *Discover Magazine*. https://www.discovermagazine.com/mind/brainwaves-in-organoids

52 Valli, K., Revonsuo, A., Pälkäs, O., Ismail, K. H., Ali, K. J., & Punamäki, R. L. (2005). The threat simulation theory of the evolutionary function of dreaming: Evidence from dreams of traumatized children. *Consciousness and Cognition, 14*(1), 188–218. https://doi.org/10.1016/S1053-8100(03)00019-9

53 Wamsley, E. J. (2014). Dreaming and offline memory consolidation. *Current Neurology and Neuroscience Reports*, *14*(3), 433. https://doi.org/10.1007/s11910-013-0433-5

54 Hobson, J. A., Hong, C. C-H., & Friston, K. J. (2014). Virtual reality and consciousness inference in dreaming. *Frontiers in Psychology*, *5*, 1133. https://doi.org/10.3389/fpsyg.2014.01133

55 Domhoff, G. W., & Fox, K. C. R. (2015). Dreaming and the default network: A review, synthesis, and counterintuitive research proposal. *Consciousness and Cognition*, *33*, 342–353. https://doi.org/10.1016/j.concog.2015.01.019

56 Hobson, J. A. (2009). REM sleep and dreaming: towards a theory of protoconsciousness. *Nature Reviews Neuroscience*, *10*, 803–813. https://doi.org/10.1038/nrn2716

57 Osaki, T., Duenki, T., Chow, S.Y.A. et al. (2024). Complex activity and short-term plasticity of human cerebral organoids reciprocally connected with axons. *Nat Commun*, *15*, 2945. https://doi.org/10.1038/s41467-024-46787-7

Chapter 10: Brain Organoids in Space: A New Frontier

58 Muotri, A. R. (2019). A Space Organoid. UC San Diego School of Medicine. https://health.ucsd.edu/news/press-releases/2019-07-08-2019-a-space-organoid/

59 Space Tango. (2023, November 9). CRS-29 Payloads. https://spacetango.com/latest/missions/crs-29-payloads/

60 Texas Children's Hospital and Baylor College of Medicine. (2024, May 3). Texas Children's Hospital and Baylor College of Medicine send human brain organoids to the International Space Station.

https://www.texaschildrens.org/content/news/texas-childrens-hospital-and-baylor-college-medicine-send-human-brain-organoids

61 Mansour, A. A., Gonçalves, J. T., Bloyd, C. W., Li, H., Fernandes, S., Quang, D., Johnston, S., Parylak, S. L., Jin, X., & Gage, F. H. (2018). An in vivo model of functional and vascularized human brain organoids. *Nature Biotechnology*, *36*(5), 432–441. https://doi.org/10.1038/nbt.4127

Chapter 11: Lab Life: The Consciousness Debate

62 Silver, D., Hubert, T., Schrittwieser, J., & Hassabis, D. (2018, December 6). *AlphaZero: Shedding new light on chess, shogi, and Go*. Google DeepMind. https://deepmind.google/discover/blog/alphazero-shedding-new-light-on-chess-shogi-and-go/

63 Internet Encyclopedia of Philosophy. (n.d.). Integrated information theory of consciousness. https://iep.utm.edu/integrated-information-theory-of-consciousness/

64 VanRullen, R., & Kanai, R. (2021). Deep learning and the Global Workspace Theory. *Trends in Neurosciences*, *44*(9), 692–704. https://doi.org/10.1016/j.tins.2021.04.005

65 Musser, G. (2024, January 17). Can quantum hints in the brain revive a radical consciousness theory? *New Scientist*, *261*(3474). https://www.newscientist.com/article/mg26134740-800-can-quantum-hints-in-the-brain-revive-a-radical-consciousness-theory/

66 Roxford, J., & Bayne, T. (2024). The case against organoid consciousness. *Neuroethics*, *17*(13). https://doi.org/10.1007/s12152-024-09548-3

67 Kosik, K. S. (2024). Why brain organoids are not conscious yet. *Perspective*, *5*(8101011). https://doi.org/10.1016/j.patter.2024.101011

68 Gaillard, M. (2024). Global versus local theories of consciousness and the consciousness assessment issue in brain organoids. *Neuroethics*, *17*(9). https://doi.org/10.1007/s12152-024-09544-7

69 Zeki, S., & Bartels, A. (1999). Toward a theory of visual consciousness. *Conscious Cognition*, *8*(2), 225–259. https://doi.org/10.1006/ccog.1999.0390

70 Hartung, T., Morales Pantoja, I. E., & Smirnova, L. (2024). Brain organoids and organoid intelligence from ethical, legal, and social points of view. *Frontiers in Artificial Intelligence*, *6*:1307613. https://doi.org/10.3389/frai.2023.1307613

71 Lavazza, A. (2021). Human cerebral organoids and consciousness: a double-edged sword. *Monash Bioethics Review, 39(2),* 129–149. https://doi.org/10.1007/s40592-021-00133-5

72 Falk, D. (2023, September 25). *Is consciousness part of the fabric of the universe?* Scientific American. https://www.scientificamerican.com/article/is-consciousness-part-of-the-fabric-of-the-universe/

73 FitMind. (2020, October 1). Self-stories: Fabrications of the mind. https://fitmind.org/blog-collection/stories-narratives-of-mind

74 Sbrobert. (2024, March 29). *Famed neuroscientist Christof Koch speaks on consciousness*. Edhat. https://www.edhat.com/news/famed-neuroscientist-christof-koch-speaks-on-consciousness/

75 Revah, O., Gore, F., Kelley, K. W., Andersen, J., Sakai, N., Chen, X., Li, M.-Y., Birey, F., Yang, X., Saw, N. L., Baker, S. W., Amin, N. D., Kulkarni, S., Mudipalli, R., Cui, B., Nishino, S., Grant, G. A., Knowles, J. K., Shamloo, M., … Pașca, S. P. (2022). Maturation and circuit integration of transplanted human cortical organoids. *Nature*, *610*, 319–326. https://doi.org/10.1038/s41586-022-05277-w

76 Erler, A. (2024). Human brain organoid transplantation: Testing the foundations of animal research ethics. *Neuroethics*, *17*(2), 1–14. https://doi.org/10.1007/s12152-023-09549-8

77 Mehta, S. A., Saharia, K. K., Nellore, A., Blumberg, E. A., & Fishman, J. A. (2023). Infection and clinical xenotransplantation: Guidance from the Infectious Disease Community of Practice of the American Society of Transplantation. *American Journal of Transplantation, 23*(3), 309–315. https://doi.org/10.1016/j.ajt.2022.12.013

Chapter 12: OI Biosecurity Concerns

78 Public Health Agency of Canada. (2018, June 29). Canadian biosafety guideline – Dual-use in life science research. Government of Canada. https://www.canada.ca/content/dam/phac-aspc/documents/programs/consultation-biosafety-guideline-dual-use-life-science-research/consultation-biosafety-guideline-dual-use-life-science-research.pdf

79 Volzhenin, K., Changeux, J.-P., & Dumas, G. (2022). Multilevel development of cognitive abilities in an artificial neural network. *Proceedings of the National Academy of Sciences, 119*(31), Article e2201304119. https://doi.org/10.1073/pnas.2201304119

80 Awan, A. A. (2023, July 21). *What is superalignment & why it is important?* KDnuggets. https://www.kdnuggets.com/2023/07/superalignment-important.html

81 National Human Genome Research Institute. (n.d.). *Review of the Ethical, Legal and Social Implications Research Program and related activities (1990-1995).* https://www.genome.gov/10001747/elsi-program-review-19901995

Chapter 13: OI A Sustainable Future

82 Wolfram, S. (2015, November 2). *George Boole: A 200-year view.* Stephen Wolfram Writings. https://writings.stephenwolfram.com/2015/11/george-boole-a-200-year-view/

83 Yamazaki, K., Vo-Ho, V. K., Bulsara, D., & Le, N. (2022). Spiking neural networks and their applications: A review. *Brain Sciences, 12*(7), 863. https://doi.org/10.3390/brainsci12070863

84 Yao, M., Richter, O., Zhao, G., Qiao, N., Xing, Y., Wang, D., Hu, T., Fang, W., Demirci, T., De Marchi, M., Deng, L., Yan, T., Nielsen, C., Sheik, S., Wu, C., Tian, Y., Xu, B., & Li, G. (2024). Spike-based dynamic computing with asynchronous sensing-computing neuromorphic chip. *Nature Communications, 15*, 4464. https://doi.org/10.1038/s41467-024-47811-6

85 Grad, P. (2023, December 12). *Brain tissue on a chip achieves voice recognition.* Tech Xplore. https://techxplore.com/news/2023-12-brain-tissue-chip-voice-recognition.html

86 Cai, H., Ao, Z., Tian, C., Wu, Z., Liu, H., Tchieu, J., Gu, M., Mackie, K., & Guo, F. (2023). Brain organoid computing for artificial intelligence. bioRxiv. https://doi.org/10.1101/2023.02.28.530502

Chapter 14: The Ethics of OI: Balancing Promise and Peril

87 Mucci, T. (2024, June 7). *What is the technological singularity?* IBM. https://www.ibm.com/think/topics/technological-singularity

Chapter 15: Brain Organoids and the Soul

88 Tubali, S. (2023, October 9). Personal communication.

89 Glass, A. (2018, August 9). *Stem cell research gets federal OK, Aug. 9, 2001.* POLITICO. https://www.politico.com/story/2018/08/09/stem-cell-research-gets-federal-ok-aug-9-2001-764402

90 Pontifical Academy for Life. (n.d.). In Wikipedia. Retrieved September 15, 2024, from https://en.wikipedia.org/wiki/Pontifical_Academy_for_Life

91 Pew Research Center. (2009, November 5). *Religion and Science in the United States.* https://www.pewresearch.org/religion/2009/11/05/an-overview-of-religion-and-science-in-the-united-states/

92 Berzin, A. (n.d.). *Right View and Intention. Study Buddhism.* Retrieved September 15, 2024, from https://studybuddhism.com/en/tibetan-buddhism/path-to-enlightenment/the-graded-path/a-full-buddhist-life-the-eightfold-path/right-view-and-intention

93 Bogle, A. (2024, September 16). Personal communication.

94 Nevins, D. (2024, September 16). Personal communication.

95 Nordin, M. M. (2024, June 22). *Islamic Bioethics of Scientific Biomedical Research (IJHHS).* Dr Musa Mohd Nordin. https://drmusanordin.com/2024/06/22/islamic-bioethics-of-scientific-biomedical-research/

Chapter 16: OI and Media

96 Presley, A., Samsa, L. A., & Dubljević, V. (2022). Media portrayal of ethical and social issues in brain organoid research. *Philosophy, Ethics, and Humanities in Medicine*, *17*(1), 8.

https://doi.org/10.1186/s13010-022-00119-z

97 Bennett Moses, L. (2015). How to think about law, regulation and technology: Problems with 'technology' as a regulatory target. *Law, Innovation and Technology*, *5*(1), 1–20. https://doi.org/10.5235/17579961.5.1.1

98 Herzfeld, N. (2017, August 10). *Noreen Herzfeld on "Artificial Intelligence: An Existential Risk?"* https://digitalcommons.csbsju.edu/theology_pubs/74/

Chapter 17: Ancient Wisdom, Modern Minds

99 O'Keefe, E. A. (1984). Towards an understanding of the significance of "The Dreamtime" to Aboriginal people. *The Aboriginal Child at School*, *12*(4), 50–56. https://doi.org/10.1017/S0310582200013407

100 Mead, H. M. (2022, June 19). *Understanding Mātauranga Māori.* E-Tangata. https://e-tangata.co.nz/comment-and-analysis/understanding-matauranga-maori/

101 Benally, H. (1992). Spiritual knowledge for a secular society: Traditional Navajo spirituality offers lessons for the nation. *Tribal College: Journal of American Indian Higher Education*, *3*(4). https://tribalcollegejournal.org/spiritual-knowledge-secular-society-traditional-navajo-spirituality-offers-lessons-nation/

102 Lévesque, F. (2014). Revisiting Inuit Qaujimajatuqangit: Inuit knowledge, culture, language, and values in Nunavut institutions since 1999. *Études/Inuit/Studies, 38*(1–2), 115–136. https://doi.org/10.7202/1028856ar

103 Pachamama Alliance. (n.d.). *Sumak Kawsay: Teachings of Indigenous Peoples*. Retrieved September 15, 2024, from https://www.pachamama.org /sumak-kawsay

104 Seven Generations International Foundation. (n.d.). *7th Generation Principle*. Retrieved September 15, 2024, from https://theseventhgenerati on.org/blog-the-seventh-generation-principle/

Chapter 18: OI in Art and Imagination

105 Jobson, C. (2013, June 13). *Beautiful Thoughts: Artist Lisa Park Manipulates Water with Her Mind*. Colossal. https://www.thisiscolossal.com /2013/06/artist-lisa-park-manipulates-water-with-her-brain/

106 Liu, A. (n.d.). *Biologically Algorithmic*. Retrieved September 15, 2024, from https://ani-liu.com/biologically-algorithmic

107 Neural Zoo. (n.d.). *Home*. Retrieved September 15, 2024, from https:// neuralzoo.com/

Chapter 19: Prudent Regulations in the OI Era

108 Varkey, B. (2021). Principles of clinical ethics and their application in clinical practice. *Medical Principles and Practice, 30*(1), 17–28. https:// doi.org/10.1159/000509119

109 Smirnova, L., Caffo, B. S., Gracias, D. H., Huang, Q., Morales Pantoja, I. E., Tang, B., Zack, D. J., Berlinicke, C. A., Boyd, J. L., Harris, T. D., Johnson, E. C., Kagan, B. J., Kahn, J., Muotri, A. R., Paulhamus, B. L., Schwamborn, J. C., Plotkin, J., Szalay, A. S., Vogelstein, J. T., … Hartung, T. (2023). Organoid intelligence (OI): the new frontier in biocomputing and intelligence-in-a-dish. *Frontiers in Science, 1*, Article 1017235. https:// doi.org/10.3389/fsci.2023.1017235

110 UNESCO. (2005, October 19). *Universal Declaration on Bioethics and Human Rights*. UNESCO. https://en.unesco.org/themes/ethics-science-and-technology/bioethics-and-human-rights

111 United Nations Office for Disarmament Affairs. (n.d.). Biological Weapons Convention. https://www.un.org/disarmament/biological-weapons/

Appendix A: OI Visionaries– Shaping the Future

112 Center for Alternatives to Animal Testing. (n.d.). Research. Johns Hopkins Bloomberg School of Public Health. https://caat.jhsph.edu/research/

113 Department of Pediatrics. (n.d.). Muotri Lab. University of California San Diego. https://pediatrics.ucsd.edu/research/faculty-labs/muotri-lab/index.html

114 University of Luxembourg. (n.d.). Luxembourg Centre for Systems Biomedicine. https://www.uni.lu/lcsb-en/

115 D., Kyrlynn. (2023, April 29). Cortical Labs secures $15 million in funding for "Brain in a Dish" DishBrain technology. *Quantum Zeitgeist*. https://quantumzeitgeist.com/cortical-labs-secures-15-million-in-funding-for-brain-in-a-dish-dishbrain-technology/

116 Kagan, B. J., Kitchen, A. C., Tran, N. T., Rollo, B., Razi, A., & Friston, K. J. (2022). The DishBrain system: A real-time synthetic biological intelligence platform. *Neuron, 110*(24), 3952–3969. https://doi.org/10.1016/j.neuron.2022.09.001

117 Kagan B. (2024, October 09). Personal communication.

118 Jordan, F. D., Kutter, M., Comby, J.-M., Brozzi, F., & Kurtys, E. (2024). Open and remotely accessible Neuroplatform for research in wetware computing. *Frontiers in Artificial Intelligence*, 7, Article 1376042. https://doi.org/10.3389/frai.2024.1376042

119 National Science Foundation. (2024, August 30). *NSF invests $14 million in bioengineered systems for ethical biocomputing*. *NSF News*. https://new.nsf.gov/news/nsf-invests-14m-bioengineered-systems-ethical-biocomputing

REFERENCES

In this reference section, you will find a carefully curated selection of articles, academic papers, press pieces, and other resources that have significantly shaped the writing and research behind this book. Each source has played a vital role in informing my perspective and deepening my understanding of the topics discussed. Whether you're looking to expand your knowledge or dive into the finer details, these accessible sources will guide you towards a richer, more nuanced understanding. Consider this your invitation to continue exploring the fascinating ideas we've touched upon.

ADDitude Editors, & Silver, L., M.D. (2023, June 29). *The Neuroscience of the ADHD Brain*. ADDitude. Medically reviewed by ADDitude's ADHD Medical Review Panel. Verified. https://www.additudemag.com/neuroscience-of-adhd-brain/

Alfonsi, V., Gorgoni, M., Giannini, A. M., De Gennaro, L., & Scarpelli, S. (2021). Investigation on neurobiological mechanisms of dreaming in the new decade. *Brain Sciences, 11*(2), 220. https://doi.org/10.3390/brainsci11020220

Ali, J. A. (2021). Islamic perspectives on organ transplantation: A continuous debate. *Religions, 12*(8), 576. https://doi.org/10.3390/rel12080576

Alzheimer's Society. *Memory loss and dementia*. Alzheimer's Society. https://www.alzheimers.org.uk/about-dementia/symptoms-and-diagnosis/symptoms/memory-loss

Arvan, M. (2023, February 6). *Can digital computers ever achieve consciousness?* Templeton Foundation. https://www.templeton.org/news/can-digital-computers-ever-achieve-consciousness

Atlas, R. M., & Dando, M. (2006). The dual-use dilemma for the life sciences: Perspectives, conundrums, and global solutions. *Biosecurity and Bioterrorism: Biodefense Strategy, Practice, and Science, 4*(3), 276–286. https://doi.org/10.1089/bsp.2006.4.276

Bai, L., Wu, Y., Li, G., Zhang, W., Zhang, H., & Su, J. (2024). AI-enabled organoids: Construction, analysis, and application. *Bioactive Materials, 31*, 525–548. https://doi.org/10.1016/j.bioactmat.2023.09.005

Belkacem, A. N., Jamil, N., Palmer, J. A., Ouhbi, S., & Chen, C. (2020). Brain computer interfaces for improving the quality of life of older adults and elderly patients. *Frontiers in Neuroscience, 14*, 692. https://doi.org/10.3389/fnins.2020.00692

Benjamin, M. H. (2022). "There Is No 'Away'": Ecological fact as Jewish theological problem. *Religions, 13*(4), 290. https://doi.org/10.3390/rel13040290

Birey, F., Andersen, J., Makinson, C. D., Islam, S., Wei, W., Huber, N., ... & Pasca, S. P. (2017). Assembly of functionally integrated human forebrain spheroids. *Nature, 545*(7652), 54–59. https://doi.org/10.1038/nature22330

Bizzarri, M., Fedeli, V., Piombarolo, A., & Angeloni, A. (2022, October). Space Biomedicine: A unique opportunity to rethink the relationships between physics and biology. *Biomedicines, 10*(10), 2633. https://doi.org/10.3390/biomedicines10102633

Blain, L. (2023, December 12). Cyborg computer with living brain organoid aces machine learning tests. *New Atlas*. https://newatlas.com/computers/hybrid-brain-organoid-computing/

Cai, H., Ao, Z., Tian, C., Wu, Z., Liu, H., Tchieu, J., ... & Guo, F. (2023). Brain organoid reservoir computing for artificial intelligence. *Nature Electronics, 6*, 1032–1039. https://doi.org/10.1038/s41928-023-01069-w

Cakir, B., Xiang, Y., Tanaka, Y., Kural, M. H., Parent, M., Kang, Y. J., ... & Kriegstein, A. R. (2019). Engineering of human brain organoids with a functional vascular-like system. *Nature Methods, 16*(11), 1169–1175. https://doi.org/10.1038/s41592-019-0586-5

Candanosa, R. M. (2023, February 28). *Could future computers run on human brain cells?* Johns Hopkins University Hub. https://hub.jhu.edu/2023/02/28/organoid-intelligence-biocomputers/

Celestis. (n.d.). *Space Medicine, Past, Present, and Future: From Dr. McCoy to the ISS*. Celestis. https://www.celestis.com/blog/space-medicine-past-present-and-future-from-dr-mccoy-to-the-iss/

Chew, L., Añonuevo, A., Knock, E. (2022). Generating cerebral organoids from human pluripotent stem cells. In Deleyrolle, L.P. (eds) *Neural Progenitor Cells. Methods in Molecular Biology, vol 2389*. Humana, New York, NY. https://doi.org/10.1007/978-1-0716-1783-0_15

Chiaradia, I., & Lancaster, M. A. (2020). Brain organoids for the study of human neurobiology at the interface of in vitro and in vivo. *Nature Neuroscience, 23*(11), 1496–1508. https://doi.org/10.1038/s41593-020-00730-3

Cinel, C., Valeriani, D., & Poli, R. (2019, January 31). Neurotechnologies for human cognitive augmentation: Current state of the art and future prospects. *Frontiers in Human Neuroscience, 13*. https://doi.org/10.3389/fnhum.2019.00013

Clarke, S. (2023). The sanctity of life as a sacred value. *Bioethics, 37*(1), 32–39. https://doi.org/10.1111/bioe.13094

Cox, D. (2023, June 5). *Are we entering the era of personalised medicine?* BBC Future. https://www.bbc.com/future/article/20230602-are-we-entering-the-era-of-personalised-medicine

Cui, D., & Wu, F. (n.d.). The influence of media use on public perceptions of artificial intelligence in China: Evidence from an online survey. *Journal of Information Development, 37*(1). https://doi.org/10.1177/0266666919893411

D'Souza, M. H., & Patel, T. R. (2020). Biodefense implications of New-World Hantaviruses. *Frontiers in Bioengineering and Biotechnology, 8*. https://doi.org/10.3389/fbioe.2020.00925

Dee, J. P. (2024). *How Boolean algebra shapes AI*. LinkedIn. https://www.linkedin.com/posts/jprdee-paul_snsinstitutions-snsdesignthinkers-designthinking-activity-7154421210062176256-SOpO

Deininger, L., Jung-Klawitter, S., Mikut, R., Richter, P., Fischer, M., Karimian-Jazi, K., Breckwoldt, M. O., Bendszus, M., Heiland, S., Kleesiek, J., Opladen, T., Kuseyri Hübschmann, O., Hübschmann, D., & Schwarz, D. (2023). An AI-based segmentation and analysis pipeline for high-field MR monitoring of cerebral organoids. *Scientific Reports, 13*, Article 21231. https://doi.org/10.1038/s41598-023-48343-7

Diaz, T. (2022, September 3). *Aerospace medicine: The next big health frontier?* MedPage Today. https://www.medpagetoday.com/opinion/second-opinions/100536

DiCorato, A. (2022, September 29). *Brain organoids replicate key events in human brain development.* Broad Institute News. https://www.broadinstitute.org/news/brain-organoids-replicate-key-events-human-brain-development

Dimitropoulos, S. (2023, February 28). *Scientists are growing mini brains in the lab. Are they conscious?* Popular Mechanics. https://www.popularmechanics.com/science/health/a42916548/are-brain-organoids-conscious/

Ding, L., Xiao, Z., Gong, X., & Peng, Y. (2022). Knowledge graphs of ethical concerns of cerebral organoids. *Cell Proliferation, 55*(8), e13239. https://doi.org/10.1111/cpr.13239

Dottori, M., & the Australian Dementia Stem Cell Consortium. (2020). If human brain organoids are the answer to understanding dementia, what are the questions? *Journal of Biomolecular Screening, 26*(5–6). https://doi.org/10.1177/1073858420912404

Dreifus, C. (2022, October 12). *Human brains are hard to study; Sergiu Pasca grows useful substitutes.* Quanta Magazine. https://www.quantamagazine.org/human-brains-are-hard-to-study-sergiu-pasca-grows-useful-substitutes-20221012/

Elci, A. (2023, September 5). *We're at the start of a revolution: Could lab brains eventually take over AI?* Euronews. https://www.euronews.com/next/2023/09/05/were-at-the-start-of-a-revolution-could-lab-brains-eventually-take-over-ai

Eiraku, M., Watanabe, K., Matsuo-Takasaki, M., Kawada, M., Yonemura, S., Matsumura, M., Wataya, T., Nishiyama, A., Muguruma, K., & Sasai, Y. (2008). Self-organized formation of polarized cortical tissues from ESCs and its active manipulation by extrinsic signals. *Cell Stem Cell, 3*(5), 519–532. https://doi.org/10.1016/j.stem.2008.09.002

Evans, N. G. (2020). Dual-use and infectious disease research. In M. Eccleston-Turner & I. Brassington (Eds.), *Infectious diseases in the new millennium* (pp. 193–215). Springer, Cham. https://doi.org/10.1007/978-3-030-39819-4_9

Eura, N., Matsui, T. K., Luginbühl, J., Matsubayashi, M., Nanaura, H., Shiota, T., … & Zheng, C. (2020). Brainstem organoids from human pluripotent stem cells. *Frontiers in Neuroscience, 14*, 538. https://doi.org/10.3389/fnins.2020.00538

Fan, S. (2023, August 8). *Heart organoids tricked out with nanowires restored heart function in rats.* Singularity Hub. https://singularityhub.com/2023/08/08/heart-organoids-tricked-out-with-nanowires-restored-heart-function-in-rats/

Fatehullah, A., Tan, S. H., & Barker, N. (2016). Organoids as an in vitro model of human development and disease. *Nature Cell Biology, 18*(3), 246–254. https://doi.org/10.1038/ncb3312

Fauzia, M. (2023, August 9). *Tiny brains in a dish could finally uncover the long-standing mysteries of psychedelics.* Inverse. https://www.inverse.com/health/tiny-brains-could-uncover-psychedelics-mysteries

FoNS News. (2023, September 4). *Accelerating the development of AI by creating artificial brain organoids.* Imperial Natural Sciences Blog. https://blogs.imperial.ac.uk/natural-sciences/2023/09/04/accelerating-the-development-of-ai-by-creating-artificial-brain-organoids/

Frank, M. (2023, October 25). *Four lessons from historical tech regulation to aid AI policymaking.* Center for Strategic & International Studies. https://www.csis.org/analysis/four-lessons-historical-tech-regulation-aid-ai-policymaking

Garreta, E., Kamm, R. D., Chuva de Sousa Lopes, S. M., Lancaster, M. A., Weiss, R., Trepat, X., Hyun, I., & Montserrat, N. (2021). Rethinking organoid technology through bioengineering. *Nature Materials, 20*, 145–155. https://doi.org/10.1038/s41563-020-00804-4

Georgalis, G. L. (2023). Evolution of consciousness. *Life, 14*(1), 48. https://doi.org/10.3390/life14010048

Goh, G. D., Lee, J. M., Goh, G. L., Huang, X., Lee, S., & Yeong, W. Y. (2023, January). Machine learning for bioelectronics on wearable and implantable devices: Challenges and potential. *Tissue Engineering Part A, 29*(1–2), 20–46. http://doi.org/10.1089/ten.tea.2022.0119

Goldman, B. (2022, October 12). *Human brain cells transplanted into rat brains hold promise for neuropsychiatric research.* Stanford Medicine News Center. https://med.stanford.edu/news/all-news/2022/10/human-rat-brain-neuron.html

Groenewegen, A. (n.d.). *Kahneman fast and slow thinking explained.* Sue Behavioural Design. https://suebehaviouraldesign.com/

Haseltine, W. A. (2023, April 28). *Forget AI; Organoid intelligence may soon power our computers.* Forbes. https://www.forbes.com/sites/williamhaseltine/2023/04/28/forget-ai-organoid-intelligence-may-soon-power-our-computers/

Haseltine, W. A. *Forget AI; Organoid intelligence may soon power our computers.* Forbes. https://www.forbes.com/sites/williamhaseltine/2023/04/28/forget-ai-organoid-intelligence-may-soon-power-our-computers/

Hyun, I., Scharf-Deering, J. C., & Lunshof, J. E. (2020). Ethical issues related to brain organoid research. *Brain Research, 1732*, 146653. https://doi.org/10.1016/j.brainres.2020.146653

Jalan-Sakrikar, N., Brevini, T., Huebert, R. C., & Sampaziotis, F. (2023). Organoids and regenerative hepatology. *Hepatology, 77*(1), 305–322. https://doi.org/10.1002/hep.32583

Kaneko, M., Narukawa, M., & Yamaguchi, S. (2021). Approval success rates of drug candidates based on target, action, modality, application, and their combinations. *Clinical and Translational Science, 14*(3), 1113–1122. https://doi.org/10.1111/cts.12980

Kelava, I., & Lancaster, M. A. (2016). Dishing out mini-brains: Current progress and future prospects in brain organoid research. *Developmental Biology, 420*(2), 199–209. https://doi.org/10.1016/j.ydbio.2016.06.037

Kemeny, R. (2016, July 28). *Minding the pulse of memory consolidation: Studying sleep spindles could help neuroscientists better understand certain cognitive impairments.* The Scientist. https://www.the-scientist.com/news-opinion/minding-the-pulse-of-memory-consolidation-33125

Lancaster, M. A., Renner, M., Martin, C. A., Wenzel, D., Bicknell, L. S., Hurles, M. E., ... & Knoblich, J. A. (2013). Cerebral organoids model human brain development and microcephaly. *Nature, 501*(7467), 373–379. https://doi.org/10.1038/nature12517

Lancaster, M. A., & Knoblich, J. A. (2014). Generation of cerebral organoids from human pluripotent stem cells. *Nature Protocols, 9*(10), 2329–2340. https://doi.org/10.1038/nprot.2014.158

Li, Y., Zeng, P. M., Wu, J., Peng, Y., Zhou, J., Zhang, Q., ... & Pei, G. (2023). *Advances and applications of brain organoids.* Neuroscience Bulletin. https://doi.org/10.1007/s12264-023-01065-2

Mansour, A. A., Gonçalves, J. T., Bloyd, C. W., Li, H., Fernandes, S., Quang, D., Johnston, S., Parylak, S. L., Jin, X., & Gage, F. H. (2018). An in vivo model of functional and vascularized human brain organoids. *Nature Biotechnology, 36*(5), 432–441. https://doi.org/10.1038/nbt.4127

Martone, R. (2019, August 27). *A successful artificial memory has been created.* Scientific American. https://www.scientificamerican.com/article/a-successful-artificial-memory-has-been-created/

McMurray, C. (2022, March 22). *How the brain senses time.* BrainFacts/SfN. https://www.brainfacts.org/thinking-sensing-and-behaving/thinking-and-awareness/2022/how-the-brain-senses-time-032222

Morris, A. (2023, January 12). *Mature 'lab grown' neurons hold promise for neurodegenerative disease*. Northwestern University Feinberg School of Medicine News Center. https://news.feinberg.northwestern.edu/2023/01/12/mature-lab-grown-neurons-hold-promise-for-neurodegenerative-disease/

Nasser, R. A., Arya, S. S., Alshehhi, K. H., Teo, J. C. M., & Pitsalidis, C. (2024, January 5). *Conducting polymer scaffolds: A new frontier in bioelectronics and bioengineering*. Trends in Biotechnology. https://doi.org/10.1016/j.tibtech.2023.11.017

Reed, J. and McFadden, J. (2024, February 4). *Neuralink: Can Musk's brain technology change the world?* BBC News. https://www.bbc.com/news/health-68169082

Ortolano, N. (2023, February 7). *Organoids bring drug discovery and development to the culture hood*. 10x Genomics Blog. https://www.10xgenomics.com/blog/organoids-bring-drug-discovery-and-development-to-the-culture-hood

Osaki, T., & Ikeuchi, Y. (2021). Advanced Complexity and Plasticity of Neural Activity in Reciprocally Connected Human Cerebral Organoids bioRxiv. https://doi.org/10.1101/2021.02.16.431387

Peter, R. M. (2024, January 12). *AI and a brain organoid: Can this combo take on supercomputers?* Labiotech.

Phillips, C. (2022). The question is not "can humans talk?" or "can they suffer?" but "can they reason?". *Animal Sentience, 31*(5). https://doi.org/10.51291/2377-7478.1704

Polti, I., Martin, B., & van Wassenhove, V. (2018). The effect of attention and working memory on the estimation of elapsed time. *Scientific Reports, 8*, 6690. https://doi.org/10.1038/s41598-018-25119-y

Prescott, T. J., Camilleri, D., Martinez-Hernandez, U., Damianou, A., & Lawrence, N. D. (2019). Memory and mental time travel in humans and social robots. *Philosophical Transactions of the Royal Society B: Biological Sciences, 374*(1771), 20180025. https://doi.org/10.1098/rstb.2018.0025

Qian, X., Song, H., & Ming, G. L. (2019). Brain organoids: Advances, applications, and challenges. *Development, 146*(8), dev166074. https://doi.org/10.1242/dev.166074

Qian, X., Nguyen, H. N., Jacob, F., Song, H., & Ming, G. L. (2017). Using brain organoids to understand Zika virus-induced microcephaly. *Development, 144*(6), 952–957. https://doi.org/10.1242/dev.140707

Randieri, C. (2023, November 15). *The Emergence of Organoid Intelligence: Reshaping AI With Miniature Brains*. Forbes. https://www.forbes.com/councils/forbestechcouncil/2023/11/15/the-emergence-of-organoid-intelligence-reshaping-ai-with-miniature-brains/

Ratliff, M., Kim, H., Qi, H., Kim, M., Ku, B., Azorin, D. D., ... & Kwon, Y.-J. (2022). Patient-derived tumor organoids for guidance of personalized drug therapies in recurrent glioblastoma. *International Journal of Molecular Sciences, 23*(12), 6572. https://doi.org/10.3390/ijms23126572

Russomano, T. (n.d.). *Space medicine – challenges and achievements*. Asian Hospital & Healthcare Management. https://www.asianhhm.com/healthcare-management/space-medicine

Safron, A. (2021). The radically embodied conscious cybernetic Bayesian brain: From free energy to free will and back again. *Entropy, 23*(6), 783. https://doi.org/10.3390/e23060783

Sanders, L. (2021, September 3). *How personalized brain organoids could help us demystify disorders*. Science News. https://www.sciencenews.org/article/brain-human-organoids-personalized-cluster-cells-study-disorders

Sawai, T., Sakaguchi, H., Thomas, E., Takahashi, J., & Fujita, M. (2019). The ethics of cerebral organoid research: Being conscious of consciousness. *Stem Cell Reports, 13*(3), 440–447. https://doi.org/10.1016/j.stemcr.2019.08.003

Shi, H., Kowalczewski, A., Vu, D., Liu, X., Salekin, A., Yang, H., & Ma, Z. (2024). Organoid intelligence: Integration of organoid technology and artificial intelligence in the new era of in vitro models. *Medicine in Novel Technology and Devices, 21*, 100276. https://doi.org/10.1016/j.medntd.2023.100276

Shou, Y., Liang, F., Xu, S., & Li, X. (2020). The application of brain organoids: From neuronal development to neurological diseases. *Frontiers in Cell and Developmental Biology, 8*, 579659. https://doi.org/10.3389/fcell.2020.579659

Silva, G. A. (2022, December 20). *Personalized neuroscience: Are human-derived brain organoids the answer?* Forbes. https://www.forbes.com/sites/gabrielasilva/2022/12/20/personalized-neuroscience-are-human-derived-brain-organoids-the-answer/

Sloan, S. A., Darmanis, S., Huber, N., Khan, T. A., Birey, F., Caneda, C., … & Pasca, S. P. (2017). Human astrocyte maturation captured in 3D cerebral cortical spheroids derived from pluripotent stem cells. *Neuron, 95*(4), 779–790. https://doi.org/10.1016/j.neuron.2017.07.035

Sozzi, M., Kajtez, J., Bruzelius, A., Wesseler, J., Nilsson, M., Birtele, M., … & Fiorenzano, A. (2022). Silk scaffolding drives self-assembly of functional and mature human brain organoids. *Nature Communications, 13*(1), 1–17. https://doi.org/10.3389/fcell.2022.1023279

Strickland, A. (2023, March 2). *'Organoid intelligence' could create brain cell-led computers.* CNN. https://edition.cnn.com/2023/03/02/world/brain-computer-organoids-scn/index.html

Sun, D., Gao, W., Hu, H., & Zhou, S. (2022, July). Why 90% of clinical drug development fails and how to improve it? *Acta Pharmaceutica Sinica B, 12*(7), 3049–3062. https://doi.org/10.1016/j.apsb.2022.02.002

Sun, N., Meng, X., Liu, Y., Song, D., Jiang, C., & Cai, J. (2021). Applications of brain organoids in neurodevelopment and neurological diseases. *Journal of Biomedical Science, 28*, 30. https://doi.org/10.1186/s12929-021-00728-4

Taraska, P. A. (2017). *How can the use of human enhancement (HE) technologies in the military be ethically assessed?* (Doctoral dissertation, Duquesne University). https://dsc.duq.edu/etd/148

Thomson, H. (2022, June 15). *How do we sense time? The brain cells that order our memories.* New Scientist. https://www.newscientist.com/article/mg25433910-700-how-do-we-sense-time-the-brain-cells-that-order-our-memories/

Tozer, L. (2023, December 11). *'Biocomputer' combines lab-grown brain tissue with electronic hardware*. Nature. https://www.nature.com/articles/d41586-023-03975-7

Tse, L. (2022, May 30). *Superposition and entanglement flee the quantum nest.* Quantum Mechanics Research Update. Physics World. https://physicsworld.com/a/superposition-and-entanglement-flee-the-quantum-nest/

Tyrell, J. (2023, May 4). *Brain-inspired computing increases the likelihood of energy-efficient AI.* TechHQ. https://techhq.com/2023/05/brain-inspired-computing-could-solve-energy-efficient-ai-puzzle/

University of Cambridge. (2021, October 21). *Lab-grown 'mini brains' hint at treatments for neurodegenerative diseases.* Retrieved from https://www.cam.ac.uk/research/news/lab-grown-mini-brains-hint-at-treatments-for-neurodegenerative-diseases

Van Norman, G. A. (2019, November). Limitations of animal studies for predicting toxicity in clinical trials: Is it time to rethink our current approach? *JACC: Basic to Translational Science, 4*(7), 845–854. https://doi.org/10.1016/j.jacbts.2019.10.008

Vetere, G., Tran, L. M., Moberg, S., Steadman, P. E., Restivo, L., Morrison, F. G., … & Frankland, P. W. (2019). Memory formation in the absence of experience. *Nature Neuroscience, 22*, 933–940. https://pubmed.ncbi.nlm.nih.gov/31036944/

Wang, H. (2018). Modeling neurological diseases with human brain organoids. *Frontiers in Synaptic Neuroscience, 10*, 15. https://doi.org/10.3389/fnsyn.2018.00015

Wang, X., Liang, Q., Luo, Y., Ye, J., Yu, Y., & Chen, F. (2024, February). Engineering the next generation of theranostic biomaterials with synthetic biology. *Bioactive Materials, 32*, 514–529. https://doi.org/10.1016/j.bioactmat.2023.10.018

Whitehead, R., Bates, G., Elphinstone, B., Yang, Y., & Murray, G. (2018). Letting go of self: The creation of the Nonattachment to Self Scale. *Frontiers in Psychology, 9*, Article 2544. https://doi.org/10.3389/fpsyg.2018.02544

Whitten, A. (2022, October 12). *Lab-grown human cells form working circuits in rat brains*. Quanta Magazine. https://www.quantamagazine.org/lab-grown-human-cells-form-working-circuits-in-rat-brains-20221012/

Wittmann, M. (2009). The inner experience of time. *Philosophical Transactions of the Royal Society B: Biological Sciences, 364*(1525), 1955–1967. https://doi.org/10.1098/rstb.2009.0003

Wulansari, N., & Darsono, W. H. W. (2021). Brain organoid: A 3D technology for investigating cellular composition and interactions in human neurological development and disease models in vitro. *Stem Cell Research & Therapy, 12*(1), 1–15. https://doi.org/10.1186/s13287-021-02369-8

Xu, J., & Wen, Z. (2021). Brain organoids: Studying human brain development and diseases in a dish. *Stem Cells International, 2021*, 5902824. https://doi.org/10.1155/2021/5902824

Ylenfest. (2018, April 5). *What can an 11th-century Islamic philosopher teach us about 21st-century neuroscience?* Petrie-Flom Center Blog. Retrieved from https://blog.petrieflom.law.harvard.edu/2018/04/05/what-can-an-11th-century-islamic-philosopher-teach-us-about-21st-century-neuroscience/

Zhang, D., Du, X., Zhang, X., Li, K., Kong, F., Cheng, G., & Zhao, S. (2020). *In vitro induction and in vivo engraftment of kidney organoids derived from human pluripotent stem cells*. Experimental and Therapeutic Medicine. **https://doi.org/10.3892/etm.2020.8844

Zhang, M., Liu, Y., & Chen, Y.-G. (2020). Generation of 3D human gastrointestinal organoids: Principles and applications. *Cell Regeneration, 9*, 6. https://doi.org/10.1186/s13619-020-00040-w

INDEX

C

O

P

Made in the USA
Monee, IL
04 November 2026

67378108R00134